# Exploring Operations Research with R

*Exploring Operations Research with R* shows how the R programming language can be a valuable tool – and way of thinking – which can be successfully applied to the field of operations research (OR). This approach is centred on the idea of the future OR professional as someone who can combine knowledge of key OR techniques (e.g., simulation, linear programming, data science, and network science) with an understanding of R, including tools for data representation, manipulation, and analysis. The core aim of the book is to provide a self-contained introduction to R (both Base R and the tidyverse) and show how this knowledge can be applied to a range of OR challenges in the domains of public health, infectious diseases, and energy generation, and thus provide a platform to develop actionable insights to support decision making.

**Features**

- Can serve as a primary textbook for a comprehensive course in R, with applications in OR
- Suitable for post-graduate students in OR and data science, with a focus on the computational perspective of OR
- The text will also be of interest to professional OR practitioners as part of their continuing professional development
- Linked to a Github repository including code, solutions, data sets, and other ancillary material.

**Jim Duggan** is a Personal Professor in Computer Science at the University of Galway, Ireland. He lectures on R, MATLAB®, and system dynamics, and he is a certified RStudio tidyverse instructor. His research interests are interdisciplinary and focus on the use of simulation and computational methods to support public health policy. You can learn more about his work on R and computation modelling on his GitHub site https://github.com/JimDuggan.

# Chapman & Hall/CRC Series in Operations Research

*Series Editors:*
*Malgorzata Sterna, Bo Chen, Michel Gendreau, and Edmund Burke*

*For more information about this series please visit: https://www.routledge.com/Chapman--HallCRC-Series-in-Operations-Research/book-series/CRCOPSRES*

# Exploring Operations Research with R

**Jim Duggan**
University of Galway, Ireland

CRC Press
Taylor & Francis Group
Boca Raton London New York

CRC Press is an imprint of the
Taylor & Francis Group, an **Informa** business

A CHAPMAN & HALL BOOK

Designed cover image: ShutterStock Images

First edition published 2024
by CRC Press
2385 NW Executive Center Drive, Suite 320, Boca Raton FL 33431

and by CRC Press
4 Park Square, Milton Park, Abingdon, Oxon, OX14 4RN

*CRC Press is an imprint of Taylor & Francis Group, LLC*

© 2024 Jim Duggan

ISBN: 978-1-032-27754-7 (hbk)
ISBN: 978-1-032-27716-5 (pbk)
ISBN: 978-1-003-29391-0 (ebk)

DOI: 10.1201/ 9781003293910

Typeset in Latin Modern font
by KnowledgeWorks Global Ltd.

*Publisher's note*: This book has been prepared from camera-ready copy provided by the authors.

For Marie, Kate, and James.

# Contents

# *Preface*

Management is the process of converting information into action.
The conversion process we call decision making.

— Jay W. Forrester (Forrester, 1961)

The central idea behind this book is that R is a valuable computational tool that can be applied to the field of operations research. R provides excellent features such as data representation, data manipulation, and data analysis. These features can be integrated with operations research techniques (e.g., simulation, linear programming and data science) to support an information workflow which can provide insights to decision makers, and so, to paraphrase the words of Jay W. Forrester, *help convert information into action.*

R is an open source programming language, with comprehensive support for mathematics and statistics. With the development of R's tidyverse — an integrated system of packages for data manipulation, exploration, and visualization — the use of R has seen significant growth, across many domains. The Comprehensive R Archive Network (CRAN) provides access to thousands of special purpose R packages (for example `ggplot2` for visualization), and these can be integrated into an analyst's workflow.

## Book structure

The book comprises three parts, where each part contains thematically related chapters:

1. *Part I* introduces R, and provides a step-by-step guide to the key features of R. The initial focus is on base R, and data structures,

including: vectors, matrices, lists, data frames, and tibbles. The building blocks of R — functions — are presented, along with important ideas including environments, functionals, and the S3 object system.

2. *Part II* presents R's tidyverse and Shiny, where the main focus is on five packages: `ggplot2`, `dplyr`, `tidyr`, `purrr`, and `shiny`, as together these provide a versatile platform for rapidly analyzing, interpreting, and visualizing data.

3. *Part III* focuses on four practical examples of using R to support operations research methods. These include exploratory data analysis, linear programming, agent-based simulation, and system dynamics.

A GitHub resource https://github.com/JimDuggan/explore_or is provided with coding examples and related resources.

## What you will learn

You will learn how to program and manipulate data in R, how to harness the power of R's tidyverse, and observe how R can be used to support problem solving within the field of operations research.

## Who should read this book

The book is primarily aimed at post-graduate students in operations research. With its coverage of R, the tidyverse, and applications in agent-based simulation and system dynamics, the book also supports continuing professional development for operations research practitioners.

As R is presented from scratch, there are no prerequisites, although prior experience of a programming language would provide useful contextual knowledge.

## What is not covered in this book

There are important topics that the book does not cover, including statistical modelling, machine learning and forecasting. The book is not intended to be used as an introductory text for operations research.

## Software needed for this book

There are just three items of software that are needed as you work your way through this book:

1. You need to install R on your computer, and R can be downloaded from CRAN at https://cran.r-project.org.

2. You need to install RStudio, which is an integrated development environment (IDE) to support R programming. RStudio can be downloaded from https://posit.co/downloads/.

3. You will have to download specific packages from CRAN that are used in the book, and this can be done from within RStudio. Examples of packages are ggplot2 to support visualization, and deSolve to support system dynamics modelling.

A screenshot of RStudio's IDE is shown in Figure 1, and this shows four different panels.

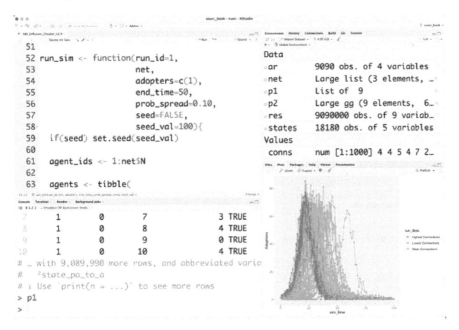

**FIGURE 1** Screenshot of RStudio

Moving anti-clockwise from the top left, these panels are:

• A source code editor, where R scripts can be created, edited, and saved.

- The R console, which allows you to create and explore variables, and observe the immediate impact of passing an instruction to R.
- Files and plots, where we can see the full file system for your written code (folders, sub-folders, files), and also view any generated plots.
- The *global environment* which shows what variables you have created, and therefore which ones can be explored and used for data processing.

Here are a number of tips you may find useful when using RStudio:

1. When coding, it is recommended to create a project to store the code. The menu option for this is "File>New Project" and a project will be housed in a folder, where you will see the project file ".Rproj" which can then be re-opened using RStudio. Using the "Files" panel, you can easily add folders to a project, and organize your scripts in a coherent way.

2. The "Source" button, which appears on the editor panel, will execute the currently active R script from start to finish.

3. The question mark is used to provide help. For example `?sample` will inform you of how the function `sample()` works.

4. The double question mark provides access to further resources. For example, if you type `??ggplot2` at the console, you can browse vignettes relating to plotting with `ggplot2`.

5. The menu option "Session>Clear Workspace" will clear all variables from your global environment. This is useful before you run a script, just to ensure that no current variable could disrupt your data processing workflow.

6. The menu option "Tools>Install Packages..." will allow you to access CRAN and install any of the available libraries.

## Recommended reading

We recommend the following books, which we have found to be valuable reference resources in writing this textbook.

- *Advanced R* by Hadley Wickham (Wickham, 2019), which presents a deep dive into R, and covers many fascinating technical topics, including object-oriented programming and metaprogramming.

- *R for Data Science* by Hadley Wickham, Mine Çetinkaya-Rundel and Garrett Grolemund (Wickham et al., 2023), aimed at data scientists to show how to perform data science with R, RStudio and the tidyverse collection of R packages.

- *Forecasting Principles and Practice* by Rob J. Hyndman and George Athanasopoulos (Hyndman and Athanasopoulos, 2021), which provides a comprehensive introduction to forecasting methods in R and also presents a valuable data processing pipeline for manipulating time series data.

- *Introduction to Statistical Learning*, by Gareth James, Daniela Witten, Trevor Hastie, and Robert Tibshirani (James et al., 2013) which demonstrates the application of the statistical learning methods in R, covering both supervised and unsupervised learning.

- *Tidy Modeling with R*, by Max Kuhn and Julia Silge (Kuhn and Silge, 2022), which introduces a collection of R packages to support modelling and machine learning.

## Acknowledgments

I would like to thank students from the University of Galway's M.Sc. in Computer Science (Data Analytics) for their enthusiasm (and valuable feedback) on the module *Programming for Data Analytics*, a module I first lectured in September 2015, and which forms the basis of Parts I and II in this textbook. Thanks to my colleagues in the School of Computer Science, and across the University of Galway, for providing a collegial environment for teaching, learning, and research; and to my colleagues in the Insight Centre for Data Analytics, the PANDEM-2 project, the SafePlan project, and the Irish Epidemiological Modelling Advisory Group, for contributing to my technical, mathematical, and public health knowledge.

Thanks to my colleagues in the communities of the System Dynamics Society and the Operational Research Society, where I had many opportunities to host workshops demonstrating how R can be used to support system dynamics and operations research. Thanks to the CRC Press/Taylor & Francis Group Editorial Team, in particular, Callum Fraser and Mansi Kabra, for providing me with the opportunity to propose, and write this textbook; and to Michele Dimont, Project Editor, for her support during the final production process.

Finally, a special thank you to my family for their encouragement, inspiration, and support.

Jim Duggan
University of Galway
Galway, Ireland
February 2024

# About the Author

Jim Duggan is a Personal Professor in Computer Science at the University of Galway, Ireland. He lectures courses in R, MATLAB, and System Dynamics, and is a certified RStudio tidyverse instructor. His research interests are interdisciplinary and focus on the use of simulation and computational methods to support public health policy. You can learn more about his work on R and computation modelling on his GitHub site https://github.com/JimDuggan. He is also author of the textbook *Systems Dynamics Modelling with R*.

# Part I

# Base R

# 1

## Getting Started with R

R is an extremely versatile open source programming language for statistics and data science.

— Norman Matloff (Matloff, 2011)

## 1.1 Introduction

This chapter introduces R, and presents a number of examples of R in action. The main idea is to demonstrate the value of using the console to rapidly create variables, perform simple operations, and display the results. Functions are introduced, and examples are shown of how they can accept data and generate outputs. The mechanism for loading libraries from the Comprehensive R Archive Network (CRAN) is presented, because an extensive set of libraries (e.g., the tidyverse) will be used in Parts II and III of the book. Finally, the chapter provides a short summary of two statistical approaches (and their relevant R functions) that will be utilized in later chapters: linear regression and correlation.

### Chapter structure

- Section 1.2 provides a glimpse of how the console can be effectively used to create, manipulate, and display variables
- Section 1.3 shows how we can call R functions to generate results.
- Section 1.4 highlights how we can leverage the work of the wider R community by installing packages to support special-purpose tasks.
- Section 1.5 introduces two statistical functions in R: lm() for linear regression, and cor() for correlation.

## 1.2 Exploring R via the RStudio console

An attractive feature of R is that it provides a console to support interactive programming, where the user can easily declare and manipulate variables. Here is an example where we assign 25 to the variable x, display it, and then use it to create a new variable y. As with all programming languages, the idea of variables is crucial, as they store data that is then manipulated using operators and/or functions.

```
# Assign 25 to  x
x <- 25
# Display x
x
#> [1] 25
# Add 21 to x and store the result y
y <- x + 25
# Display y
y
#> [1] 50
```

## 1.3 Calling functions

Functions are central to programming in R, and there are two kinds of functions: (1) those available within base R, and other libraries and (2) functions written by programmers for their own use. (Note the term base R is used to describe the core version of R). All functions take a set of arguments as input, and then return a result. Results from functions are then assigned to variables using the assignment operator <-. Once a variable is created, it is stored in R's *global environment*, which will be introduced in Chapter 4. We will now explore examples of using functions to create, and process, data.

The first function we use is c(), which is used to create a variable v to store a range of values. This type of data is known as an *atomic vector*, and one or many values can be stored, where all the values have the same type (for example, integer, double, logical, or character). Here we pass in three arguments, and these are then stored in successive elements of the vector v.

```
# Call the function c()
# Store the result in the variable v
v <- c(10, 20, 30)
# Display v
```

```
v
#> [1] 10 20 30
```

We can now use other R functions to process the variable v, for example, the functions `sum()`, `mean()`, and `sqrt()`. The results are shown below.

```
sum(v)
#> [1] 60
mean(v)
#> [1] 20
sqrt(v)
#> [1] 3.162 4.472 5.477
```

A full description of R's data structures will be presented in Chapters 2, 3, and 5, while functions will be introduced in Chapter 4.

## 1.4 Installing packages

A valuable feature of R is the ease with which new packages can be added, via CRAN. Packages contain data and functions, and can be installed using the function `install.packages()`, or by highlighting the "Tools>Install Packages" option in RStudio. Here we install the data package `aimsir17`, which contains Irish weather and energy grid data from 2017.

```
install.packages("aimsir17")
```

Once a package is installed, it can then be loaded for use using the `library()` function. In this case, we load `aimsir17` and then show the information that is contained in variable `stations`. The first six rows are displayed, as this is made possible by calling the function `head()`.

```
library(aimsir17)
head(stations)
#> # A tibble: 6 x 5
#>   station       county  height latitude longitude
#>   <chr>         <chr>    <dbl>    <dbl>     <dbl>
#> 1 ATHENRY       Galway      40     53.3     -8.79
#> 2 BALLYHAISE    Cavan       78     54.1     -7.31
#> 3 BELMULLET     Mayo         9     54.2    -10.0
#> 4 CASEMENT      Dublin      91     53.3     -6.44
#> 5 CLAREMORRIS   Mayo        68     53.7     -8.99
#> 6 CORK AIRPORT  Cork       155     51.8     -8.49
```

## 1.5    Using statistical functions in R

While the focus of the book is primarily on operational research methods, two statistical methods are used in a number of chapters. These are linear regression and correlation.

### 1.5.1    Linear regression

Regression analysis provides a way for predicting an outcome variable from either one input variable (simple regression) or several predictor variables (multiple regression), and the process involves generating a model that best describes the data (Field et al., 2012). The solution process for simple regression is to search for two parameters (the regression coefficients) that describe the line of best fit, which are $\beta_0$ (the intercept) and $\beta_1$ (the slope). There is also a residual term known as $\epsilon$, which represents the difference between a model prediction and the actual data, and the overall linear regression equation is defined as $Y_i = \beta_0 + \beta_1 X_i + \epsilon_i$, for each observation $i$ in a dataset (Field et al., 2012).

While a detailed presentation of the linear regression process is outside the scope of this book, we will use it later to show how straightforward it is to integrate regression models into the R data processing pipeline. We will utilize the R function `lm()`, which is used to fit linear models. Before exploring a simple example, we will provide an overview of the inputs and outputs of the regression process.

Figure 1.1 shows a scatterplot between two variables from the R dataset `mtcars`, based on 1974 motoring data relating fuel consumption and aspects of automobile design and performance. In Chapter 7 we will see how to plot this graph, while in Chapter 5 we will show how the underlying data is stored. A number of points can be made in relation to the graph:

- The variable on the x-axis is displacement, a measure of the engine size (cubic inches).
- The variable on the y-axis is miles per gallon, a measure of fuel efficiency.
- The plot highlights a negative relationship between the two variables, as an increase in displacement is linked with a decrease in miles per gallon.
- The line shown is *the line of best fit*, which seeks to minimize the overall distance between the model and the data points.

The `lm()` function will calculate the coefficients for the line of best fit for an input set of data $X_i$ and an output set of data $Y_i$. In our example, we set the variable x to store the displacement values, and y stores the corresponding miles per gallon values. Note that we will explain terms such as `mtcars$disp`

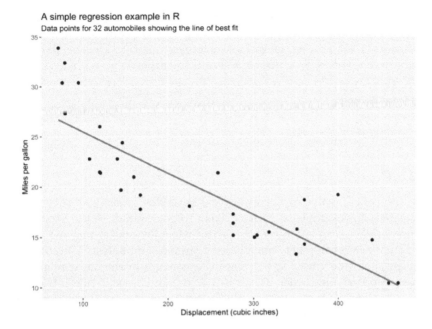

A simple regression example in R
Data points for 32 automobiles showing the line of best fit

**FIGURE 1.1** A linear model between engine size and miles per gallon (source datasets::mtcars)

later in Chapters 3 and 5, and for the moment we just assume that the terms allow us to access the required data.

We execute these two lines of code, and show the 32 observations for each variable. For example, we can see the first observation in X is 160.0 (which represents the displacement of the first car), and the corresponding value in Y is 21.0 (the miles per gallon of the first car).

```
X <- mtcars$disp
X
#>  [1] 160.0 160.0 108.0 258.0 360.0 225.0 360.0 146.7 140.8 167.6
#> [11] 167.6 275.8 275.8 275.8 472.0 460.0 440.0  78.7  75.7  71.1
#> [21] 120.1 318.0 304.0 350.0 400.0  79.0 120.3  95.1 351.0 145.0
#> [31] 301.0 121.0
Y <- mtcars$mpg
Y
#>  [1] 21.0 21.0 22.8 21.4 18.7 18.1 14.3 24.4 22.8 19.2 17.8 16.4
#> [13] 17.3 15.2 10.4 10.4 14.7 32.4 30.4 33.9 21.5 15.5 15.2 13.3
#> [25] 19.2 27.3 26.0 30.4 15.8 19.7 15.0 21.4
```

Next, we invoke the `lm()` function by passing in the regression term `Y~X` (a formula in R), and then we store results in the variable `mod`. The function `coefficents()` will process the variable `mod` to extract the two fitted parameters (which are the $\beta_0$ and $\beta_1$ terms we referred to earlier.)

```
mod <- lm(Y~X)
coefficients(mod)
#> (Intercept)          X
#>    29.59985    -0.04122
```

### 1.5.2  Correlation

Often problems involve exploring several variables to see how they may be interrelated. For example, is it more likely to get windier as the atmospheric pressure falls? A correlation problem arises when an analyst may ask whether there is any relationship between a pair of variables of interest. The correlation coefficient $r$ is widely used to calculate the strength of the linear relationship between two variables, with $-1 \leq r \leq 1$ (Hoel and Jessen, 1971). If $r$ equals $-1$ or $+1$, then all the points on a scatterplot of the two variables will lie on a straight line. The interpretation of $r$ is a purely mathematical one and completely devoid of any cause of effect implications (Hoel and Jessen, 1971). Or to use the well-known quote "correlation does not imply causation".

In R, the correlation coefficient (the default is Pearson's method) is calculated between two or more data streams using the function `cor()`. We can return to the displacement and miles per gallon data used in the previous example, and calculate the correlation coefficient between the variables. The value shows a strong negative (mathematical) relationship between the two variables.

```
cor(Y,X)
#> [1] -0.8476
```

Overall, the correlation measure is valuable during the exploratory data analysis phase, and examples of correlation calculations will be presented in Chapter 12.

## 1.6   Next steps

We are now ready to explore R in more detail and will start in Chapter 2 by introducing our first R data structure, known as a vector.

# 2

## *Vectors*

The vector type is really the heart of R. It's hard to imagine R code, or even an interactive R session, that doesn't involve vectors.

— Norman Matloff (Matloff, 2011)

## 2.1 Introduction

Knowledge of vectors is fundamental in R. A vector is a one-dimensional data structure, and there are two types of vectors: **atomic vectors**, where all the data must be of the same type, and **lists**, which are more flexible and each element's type can vary. Upon completing the chapter, you should understand:

- The difference between an atomic vector and a list, and be able to create atomic vectors and lists using the `c()` and `list()` functions.
- The four main types of atomic vector, and how different vector elements can be named.
- The rules of coercion for atomic vectors, and the importance of the function `is.na()`.
- The idea of vectorization, and how arithmetic and logical operators can be applied to vectors.
- R functions that allow you to manipulate vectors.
- How to solve all five test exercises

### Chapter structure

- Section 2.2 introduces the four main types of atomic vectors, and describes: how to create larger vectors; a mechanism known as coercion; how to name vector elements; and, how to deal with missing values.
- Section 2.3 explains vectorization, which allows for the same operation to be carried out on each vector element.
- Section 2.4 describes the list, shows how to create and name lists, and also how to create larger lists.
- Section 2.5 presents a *mini-case*, which highlights how atomic vectors can be used to simulate the throwing of two dice, and then gather frequency data on each throw. These simulated results are then compared with the expected probability values. The mini-case shows the utility of using vectors, and how R functions can speed up the analysis process.
- Section 2.6 provides a summary of all the functions introduced in the chapter.
- Section 2.7 provides a number of short coding exercises to test your understanding of the material covered.

## 2.2 Atomic vectors

There are a number of data structures in R, and the first one we explore is the *atomic vector*. This is a one-dimensional data structure that allows you to store one or more values. Note that unlike other programming languages, there is no special variable in R for storing a single value. A single variable (often called a *scalar* in other languages) in R is simply an atomic vector of size 1. An important constraint of atomic vectors is that all of the elements must be of the same type.

### 2.2.1 Primary types of atomic vectors

In order to create an atomic vector, the function `c()` is used (short for combine). When assigning a value to a variable, R's *right-to-left* assignment operator `<-` should be used, even though the operator `=` will also work. There are four main types of atomic vectors we will use [1]. In introducing each vector type, we utilize several R functions that provide additional information on each atomic vector. These are:

- `typeof()`, which shows the variable type, which will be one of the four categories.

---

[1]Note there are two other data types in R, one for complex numbers, the other for raw bytes

- `str()`, which compactly displays the internal structure of a variable, and also shows the type. This is a valuable function that you will make extensive use of in R, particularly when you are exploring data returned from functions.
- `is.logical()`, `is.double()`, `is.integer()`, and `is.character()` which tests the variable's type, and returns the logical type TRUE if the type aligns with the function name.

The four main data types are:

**logical**, where values can be either TRUE or FALSE, and the abbreviations T and F can also be used. For example, here we declare a logical vector with five elements.

```
# Create and display a logical vector
x_logi <- c(TRUE, T, FALSE, TRUE, F)
x_logi
#> [1]   TRUE   TRUE FALSE   TRUE FALSE

typeof(x_logi)
#> [1] "logical"

str(x_logi)
#>   logi [1:5] TRUE TRUE FALSE TRUE FALSE

is.logical(x_logi)
#> [1] TRUE
```

**integer**, which represents whole numbers (negative and positive), and must be declared by appending the letter L to the number. The significance of L is that it is an abbreviation for the word *long*, which is a type of integer.

```
# Create and display an integer vector
x_int <- c(2L, 4L, 6L, 8L, 10L)
x_int
#> [1]  2  4  6  8 10

typeof(x_int)
#> [1] "integer"

str(x_int)
#>   int [1:5] 2 4 6 8 10

is.integer(x_int)
#> [1] TRUE
```

**double**, which represents floating point numbers. Note that integer and double vectors are also known as numeric vectors (Wickham, 2019).

```
# Create and display a double vector
x_dbl<- c(1.2, 3.4, 7.2, 11.1, 12.7)
x_dbl
#> [1]   1.2   3.4   7.2 11.1 12.7

typeof(x_dbl)
#> [1] "double"

str(x_dbl)
#>   num [1:5] 1.2 3.4 7.2 11.1 12.7

is.double(x_dbl)
#> [1] TRUE
```

**character**, which represents values that are in string format.

```
# Create and display a character vector
x_chr<- c("One","Two","Three","Four","Five")
x_chr
#> [1] "One"   "Two"   "Three" "Four"  "Five"

typeof(x_chr)
#> [1] "character"

str(x_chr)
#>   chr [1:5] "One" "Two" "Three" "Four" "Five"

is.character(x_chr)
#> [1] TRUE
```

A further feature of the c() function is that a number of atomic vectors can be appended by including their variable names inside the function call. This is shown below, where we combine the variables v1 and v2 to generate new variables v3 and v4.

```
# Create vector 1
v1 <- c(1,2,3)
# Create vector 2
v2 <- c(4,5,6)

# Append vectors to create vector 3
v3 <- c(v1, v2)
v3
#> [1] 1 2 3 4 5 6

# Append vectors to create vector 4
```

```
v4 <- c(v2, v1)
v4
#> [1] 4 5 6 1 2 3
```

## 2.2.2 Creating larger vectors

While the c() function creates atomic vectors, there may be cases where
a larger atomic vector needs to be constructed, and there are a number of
additional ways that can be used for this.

- The colon operator : which generates a regular sequence of integers as an
  atomic vector, from a starting value to the end of a sequence. The function
  length() can also be used to confirm the number of elements in the atomic
  vector.

```
x <- 1:10
x
#>  [1]  1  2  3  4  5  6  7  8  9 10
```

```
typeof(x)
#> [1] "integer"
length(x)
#> [1] 10
```

- The sequence function seq(), which generates a regular sequence from a
  starting value (from) to a final value (to) and also allows for an increment
  between each value (by), or for a fixed length for the vector to be specified
  (length.out). Note that when calling a function in R such as the seq function,
  the argument name can be used when passing a value. This is convenient, as
  it means we don't need to know that exact positioning of the argument in
  the parameter list. This will be discussed in more detail in Chapter 4.

```
x0 <- seq(1,10)
x0
#>  [1]  1  2  3  4  5  6  7  8  9 10
```

```
x1 <- seq(from=1, to=10)
x1
#>  [1]  1  2  3  4  5  6  7  8  9 10
```

```
x2 <- seq(from=1, to=5, by=.5)
x2
#> [1] 1.0 1.5 2.0 2.5 3.0 3.5 4.0 4.5 5.0
```

```
x3 <- seq(from=1, to=10, by=2)
x3
```

```
#> [1] 1 3 5 7 9

x4 <- seq(from=1, length.out=10)
x4
#>  [1]  1  2  3  4  5  6  7  8  9 10

x5 <- seq(length.out=10)
x5
#>  [1]  1  2  3  4  5  6  7  8  9 10
```

- The replication function `rep()` replicates values contained in the input vector, a given number of times.

```
# use rep to initialize a vector to zeros
z1 <- rep(0,10)
z1
#>  [1] 0 0 0 0 0 0 0 0 0 0

# use rep to initialize with a pattern of values
z2 <- rep(c("A","B"),5)
z2
#>  [1] "A" "B" "A" "B" "A" "B" "A" "B" "A" "B"
```

- The `vector()` function creates a vector of a fixed length. This is advantageous for creating larger vectors in advance of carrying out processing operations on each individual element. This function also initializes each vector element to a default value.

```
y1 <- vector("logical",   length = 3)
y1
#> [1] FALSE FALSE FALSE

y2 <- vector("integer",   length = 3)
y2
#> [1] 0 0 0

y3 <- vector("double",    length = 3)
y3
#> [1] 0 0 0

y4 <- vector("character", length = 3)
y4
#> [1] "" "" ""
```

## 2.2.3 The rules of coercion

All the elements of an atomic vector must be of the same type. R will ensure that this is the case, and it enforces this through a process known as coercion. The idea is that if types are mixed in an atomic vector, R will coerce to the most *flexible type*. For example, the number 1 (integer) could also be represented as 1.0 (double). The logical type also provides an interesting example when combined with integer or double types, as TRUE values are coerced to the value 1, and FALSE values are coerced to zero.

The coercion rules are shown in the table, and several examples are provided.

|            | logical   | integer   | double    | character |
| ---------- | --------- | --------- | --------- | --------- |
| **logical**    | logical   | integer   | double    | character |
| **integer**    | integer   | integer   | double    | character |
| **double**     | double    | double    | double    | character |
| **character**  | character | character | character | character |

```
# Create a vector with logical and integer combined
ex1 <- c(T,F,T,7L)
ex1
#> [1] 1 0 1 7
typeof(ex1)
#> [1] "integer"

# Create a vector with logical and double combined
ex2 <- c(T,F,T,7.3)
ex2
#> [1] 1.0 0.0 1.0 7.3
typeof(ex2)
#> [1] "double"

# Create a vector with integer and double combined
ex3 <- c(1L,2L, 3L, 4.1)
ex3
#> [1] 1.0 2.0 3.0 4.1
typeof(ex3)
#> [1] "double"

# Create a vector with logical, integer, double and character
# combined
ex4 <- c(TRUE,1L,2.0, "Hello")
ex4
#> [1] "TRUE"  "1"     "2"     "Hello"
```

```
typeof(ex4)
#> [1] "character"
```

### 2.2.4   Naming atomic vector elements

An excellent feature of R is that vector elements (both atomic vectors and
lists) can be named, and we will see, in Chapter 3, that this feature can be
used to perform subsetting operations on vectors. Names are an *attribute* of
vectors and lists, and we will explore attributes further in Chapter 6. There
are a number of ways to name vector elements, and the simplest is to declare
the name of each element as part of the c() function, using the = symbol. Here
is an example, based on one of the earlier vectors we defined. Note that the
names do not have any impact on manipulating the data, as the summary()
function will still return the summary values for the data contained in the
vector.

```
# Create a double vector with named elements
x_dbl<- c(a=1.2, b=3.4, c=7.2, d=11.1, e=12.7)

x_dbl
#>    a    b    c    d    e
#>  1.2  3.4  7.2 11.1 12.7

summary(x_dbl)
#>    Min. 1st Qu.  Median    Mean 3rd Qu.    Max.
#>    1.20    3.40    7.20    7.12   11.10   12.70
```

A character vector of the element names can be easily extracted using a special
R function called names(), and this function also has a number of interesting
features. First, we can see how the element names are extracted.

```
# Extract the names of the x_dbl vector
x_dbl_names <- names(x_dbl)

typeof(x_dbl_names)
#> [1] "character"

x_dbl_names
#> [1] "a" "b" "c" "d" "e"
```

What is interesting about the function names() is that it can act as an *accessor
function*, that will return the names of the vector elements, and it can be
used to set the names of a vector. We can show this on an unnamed vector as
follows. Note, R will not object if you call two different elements by the same
name, although you should avoid this, as it will impact the subsetting process.

```
# Show our previously defined vector x_logi
x_logi
#> [1]   TRUE   TRUE FALSE   TRUE FALSE

# Allocal names to each vector element
names(x_logi) <- c("f","g","h","i","j")
x_logi
#>     f     g     h     i     j
#>  TRUE  TRUE FALSE  TRUE FALSE
```

### 2.2.5    Missing values - introducing NA

When working with real-world data, it is common that there will be missing values. For example, a thermometer might break down on any given day, causing an hourly temperature measurement to be missed. In R, NA is a logical constant of length one which contains a missing value indicator. Therefore, any value of a vector could have the value NA, and we can demonstrate this as follows.

```
# define a vector v
v <- 1:10
v
#> [1]  1  2  3  4  5  6  7  8  9 10

# Simulate a missing value by setting the final value to NA
v[10] <- NA
v
#> [1]  1  2  3  4  5  6  7  8  9 NA

# Notice how summary() deals with the NA value
summary(v)
#>    Min. 1st Qu.  Median    Mean 3rd Qu.    Max.    NA's
#>       1       3       5       5       7       9       1

# Notice what happens when we try to get the maximum value of v
max(v)
#> [1] NA
```

There are two follow-on points of interest. First, how can we check whether NA is present in a vector? To do this, the function is.na() must be used, and this function returns a logical vector showing whether an NA value is present at a given location. When we explore subsetting in Chapter 3, we will see how to further manipulate any missing values.

```
v
#> [1]  1  2  3  4  5  6  7  8  9 NA
```

```
# Look for missing values in the vector v
is.na(v)
#>  [1] FALSE FALSE FALSE FALSE FALSE FALSE FALSE FALSE FALSE   TRUE
```

The second point to note is to reexamine the function call `max(v)`, which returned the value `NA`. This is important, as it shows the presence of an `NA` value causes difficulty for what should be straightforward calculation. This obstacle can be overcome by adding an additional argument to the `max()` function which is `na.rm=TRUE`, which requests that `max()` omits `NA` values as part of the data processing. This option is also available for many R functions that operate on vectors, for example, `sum()`, `mean()` and `min()`.

```
v
#>  [1]  1  2  3  4  5  6  7  8  9 NA
max(v, na.rm = TRUE)
#>  [1] 9
```

## 2.3   Vectorization

Vectorization is a powerful R feature that enables a function to operate on all the elements of an atomic vector, and return the results in new atomic vector, of the same size. In these scenarios, vectorization removes the requirement to write loop structures to iterate over the entire vector, and so it leads to a simplified data analysis process.

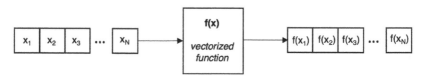

**FIGURE 2.1** Vectorization in R

The idea of vectorization is shown in Figure 2.1, and it is an intuitive concept, where the main elements are:

- an input atomic vector of size $N$ with elements $x_1, x_2, x_3, ..., x_N$,

- a vectorized function $f(x)$ that will (1) create an output atomic vector of size $N$ of the appropriate type (it does not have to be the same type as the input), and (2) loop through each element, apply the function's logic, and store the result in the output vector $f(x_1), f(x_2), f(x_3), ..., f(x_N)$.

Many R functions are vectorized, which means that they perform the same operation on each data element, and store the result in a new vector. The function `sqrt()` is a good example of a vectorized function, as it takes in a vector of numbers, and calculates the square root of each individual number.

The function `set.seed()` allows you to create the same sequence of random numbers, and we will make use of this function throughout the textbook, in order that you can replicate all the results. To generate a sample of random numbers, we use the R function `sample()`. This accepts a vector of one or more elements from which to choose, the number of random elements to generate, whether or not to use replacement, and it generates a vector of elements as output. Setting the argument `replacement=TRUE` within the call to `sample()` means that the same value can be drawn more than once, and by default, this argument set to false.

```
# Set the random number seed to 100
set.seed(100)
# Create a sample of 5 numbers from 1-10.
# Numbers can only be selected once (no replacement)
v <- sample(1:10,5)
v
#> [1] 10  7  6  3  1
length(v)
#> [1] 5
typeof(v)
#> [1] "integer"


# Call the vectorized function sqrt (square root)
rv <- sqrt(v)
rv
#> [1] 3.162 2.646 2.449 1.732 1.000
length(rv)
#> [1] 5
typeof(rv)
#> [1] "double"
```

To further explore how vectorization works, we provide three different examples: arithmetic operators, logical operators and the vectorized `ifelse` function.

## 2.3.1 Operators

Similar to other programming languages, R has arithmetic, relational, and logical operators. R's main arithmetic operators are shown below.

| R Arithmetic Operator | Description |
| --- | --- |
| + | Addition |
| – | Subtraction |
| * | Multiplication |

| R Arithmetic Operator | Description |
| --- | --- |
| / | Division |
| %/% | Integer division |
| ** or ^ | Exponentiation |
| %% | Modulus |

All of these operations support vectorization, as illustrated by the following examples.

```
# Define two sample vectors, v1 and v2
v1 <- c(10, 20, 30)
v1
#> [1] 10 20 30
v2 <- c(2, 4, 3)
v2
#> [1] 2 4 3

# Adding two vectors together
v1 + v2
#> [1] 12 24 33

# Vector subtraction
v1 - v2
#> [1]  8 16 27

# Vector multiplication
v1 * v2
#> [1] 20 80 90

# Vector division
v1 / v2
#> [1]  5  5 10

# Vector exponentiation
v1 ^ v2
#> [1]    100 160000  27000

# Integer division, returns a whole number
v1 %/% 7
#> [1] 1 2 4

# Modulus, returns the remainder
v1 %% 7
#> [1] 3 6 2
```

Note, in cases where two vectors are of unequal length, R has a *recycling* mechanism, where the shorter vector will be recycled in order to match the longer vector.

```
# Define two unequal vectors
v3 <- c(12, 16, 20, 24)
v3
#> [1] 12 16 20 24
v4 <- c(2,4)
v4
#> [1] 2 4


# Recycling addition and subtraction
v3 + v4
#> [1] 14 20 22 28


v3 - v4
#> [1] 10 12 18 20


# Recycling multiplication, division and exponentiation
v3 * v4
#> [1] 24 64 40 96


v3 / v4
#> [1]  6  4 10  6


v3 ^ v4
#> [1]    144  65536    400 331776
```

Relational operators allow for comparison between two values, and they always return a logical vector. There are six categories of relational operators, as shown in the table.

| R Relational Operator | Description |
| --- | --- |
| < | Less than |
| <= | Less than or equal to |
| > | Greater than |
| >= | Greater than or equal to |
| == | Equal to |
| != | Not equal to |

In a similar way to arithmetic operators, relational operators support vectorization. The output from a relational operator is a logical atomic vector that is the same size as the input vector, and every value contains the result of the logical operation applied to the corresponding element.

```
# Setup a test vector
v5 <- c(5,1,4,2,6,8)
v5
#> [1] 5 1 4 2 6 8

# Test for all six relational operators
v5 < 4
#> [1] FALSE   TRUE FALSE   TRUE FALSE FALSE

v5 <= 4
#> [1] FALSE   TRUE   TRUE   TRUE FALSE FALSE

v5 > 4
#> [1]   TRUE FALSE FALSE FALSE   TRUE   TRUE

v5 >= 4
#> [1]   TRUE FALSE   TRUE FALSE   TRUE   TRUE

v5 == 4
#> [1] FALSE FALSE   TRUE FALSE FALSE FALSE

v5 != 4
#> [1]   TRUE   TRUE FALSE   TRUE   TRUE   TRUE
```

The output from relational operations can take advantage of R's coercion rules. The sum() function will coerce logical values to either ones or zeros. Therefore, it is easy to find the number of values that match the relational expression. For example, in the following stream of numbers, we can see how many vector elements are greater than the mean.

```
# Setup a test vector, in this case, a sequence
v6 <- 1:10
v6
#>  [1]  1  2  3  4  5  6  7  8  9 10

# create a logical test and see the results
l_test <- v6 > mean(v6)
l_test
#>  [1] FALSE FALSE FALSE FALSE FALSE   TRUE   TRUE   TRUE   TRUE   TRUE

# Send the output to sum to see how many have matched
sum(l_test)
#> [1] 5
```

Logical operators perform operations such as AND and NOT, and for atomic vectors, the following logical operators can be used.

| R Logical Operator | Description |
| --- | --- |
| ! (NOT) | Converts TRUE to FALSE, or FALSE to TRUE |
| & (AND) | TRUE if all relational expressions are TRUE, otherwise FALSE |
| \| (OR) | TRUE if any relational expression is TRUE, otherwise FALSE |

```
# Setup a test vector, in this case, a sequence of random numbers
set.seed(200)
v  <- sample(1:20, 10, replace = T)
v
#>  [1]  6 18 15  8 12 18 12 20  8  4

# Use logical AND to see which values are in the range 10-14
v >= 10 & v <= 14
#>  [1] FALSE FALSE FALSE FALSE  TRUE FALSE  TRUE FALSE FALSE FALSE

# Use logical OR to see which values are < 5  or > 17
v < 5 | v > 17
#>  [1] FALSE  TRUE FALSE FALSE FALSE  TRUE FALSE  TRUE FALSE  TRUE

# Use logical NOT to see which values are not even
!(v %% 2 == 0)
#>  [1] FALSE FALSE  TRUE FALSE FALSE FALSE FALSE FALSE FALSE FALSE
```

### 2.3.2 ifelse() vectorization

The `ifelse()` function allows for successive elements of an atomic vector to be processed with the same test condition. The general form of the function is `ifelse(test_condition, true_value, false_value)`, where:

- `test_condition` is a logical vector, or an operation that yields a logical vector, such as a logical operator.
- `true_value` is the new vector value if the condition is true.
- `false_value` is the new vector value if the condition is false.

In the following example we have a vector sequence from 1 to 10. We calculate the mean and create a second vector that contains the value "GT" if the value is greater than the mean, and otherwise "LE". The call to `ifelse()` will create a similar size vector, where each element contains new information on the corresponding element in the original vector.

```
# Create a vector of numbers from 1 to 10
v <- 1:10
```

```
v
#>   [1]  1  2  3  4  5  6  7  8  9 10

# Calculate the mean
m_v <- mean(v)
m_v
#> [1] 5.5

# Create a new vector des_v based on a condition, and using ifelse()
des_v <- ifelse(v > m_v, "GT", "LE")
des_v
#>   [1] "LE" "LE" "LE" "LE" "LE" "GT" "GT" "GT" "GT" "GT"
```

The function ifelse() implements a looping action, as it iterates through
the entire vector and generates a vector of equal length as output. Another
example focuses on categorizing ages into *age buckets*, for example, people
whose age is less than 18 are classified as children, people greater than or equal
to 18 and less than 65 are adults, and those 65 or over are elderly. We can
achieve this classification using a nested ifelse() statement.

```
set.seed(400)
# Create a random sample of people's ages from age 1 to 90
ages <- sample(1:90,10)
ages
#>   [1] 23 87 86 28 14 13 88 71 41 45

# Create a vector that places each age into an age bucket
# Note the use of a nested ifelse statement
age_buckets <- ifelse(ages<18,"Child",
                    ifelse(ages<65,"Adult","Elderly"))
age_buckets
#>   [1] "Adult"   "Elderly" "Elderly" "Adult"   "Child"   "Child"
#>   [7] "Elderly" "Elderly" "Adult"   "Adult"
```

## 2.4  Lists

A list is a vector that can contain different types, including a list. It is a
flexible data structure and is often used to return data from a function. A
good example is the linear regression function in R, lm(), which returns a list
containing all information relating to the results of a linear regression task.
A list can be defined using the list() function, which is similar to the c()
function used to create atomic vectors. Here is an example of defining a list.

```
# Create a list
l1 <- list(1:2,c(TRUE, FALSE),list(3:4,5:6))
# Display the list.
l1
#> [[1]]
#> [1] 1 2
#>
#> [[2]]
#> [1]  TRUE FALSE
#>
#> [[3]]
#> [[3]][[1]]
#> [1] 3 4
#>
#> [[3]][[2]]
#> [1] 5 6
# Show the list type
typeof(l1)
#> [1] "list"
# Summarize the list structure
str(l1)
#> List of 3
#>  $ : int [1:2] 1 2
#>  $ : logi [1:2] TRUE FALSE
#>  $ :List of 2
#>   ..$ : int [1:2] 3 4
#>   ..$ : int [1:2] 5 6
# Confirm the number of elements
length(l1)
#> [1] 3
```

The variable l1 is a list, and it contains three elements:

- An atomic vector of two integers, as defined by the command 1:2.
- An atomic vector of two logicals, defined by c(TRUE, FALSE).
- A list containing two elements, both integer atomic vectors, the first defined by 3:4, and the second defined by 5:6.

Notice that some of the functions we used to explore atomic vectors can be used for lists. While displaying the variable itself provides information (this will be explored in more detail in Chapter 3), a valuable function for exploring lists is str(), as it provides a more succinct summary of the list size, and the structure of each element.

### 2.4.1   Visualizing lists

While atomic vectors can be visualized, for example in Figure 2.1, as a rectangular set of elements, we use a rounded rectangle to visualize a list. The list l1 is visualized in 2.2.

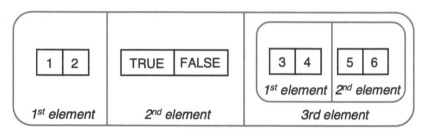

**FIGURE 2.2** Visualizing a list of three elements

The elements are separated by vertical lines, and the diagram shows the three list elements. The first is a vector of two integer elements, the second is the vector with two logical elements. The third element is interesting, and shows the flexibility of the list structure in R. This third element is a list itself, of size 2, and so it is represented by a rounded rectangle. It has its own contents, the first element being an integer vector of size 2, the second another integer vector of size 2. This example highlights an important point: the list can contain elements of different types, including another list, and is highly flexible.

### 2.4.2   Creating larger lists

As we have seen with atomic vectors, it can be valuable to create a larger structure that could then be processed using a loop. With lists, the function vector(), which we have seen used for creating atomic vectors, can also be used to create a list of a given size.

```
# Create a list
l2 <- vector(mode="list",length=3)
# Summarize the list structure
str(l2)
#> List of 3
#>  $ : NULL
#>  $ : NULL
#>  $ : NULL
# Confirm the number of elements
length(l2)
#> [1] 3
```

In this case, the variable l2 is a list that is *ready for use*, and the NULL value displayed is a reserved word in R, and is returned by expressions whose value is undefined. We will explore how to make use of this type of structure in Chapter 3.

### 2.4.3  Naming list elements

Similar to atomic vectors, each element of a list can be named. This is frequently used in R, for reasons that will become clear in Chapter 3. The main point is that naming list elements is a good idea, and can lead to (1) increased clarity when using the str() function and (2) increased flexibility when accessing a list element. Below, we take the same list l1 from before, and this time each list element is named. This naming is also applied to the "list within the list".

```
# Create a list
l1 <- list(el1=1:2,
          el2=c(TRUE, FALSE),
          el3=list(el3_el1=3:4,el3_el2=5:6))
# Summarize the list structure
str(l1)
#> List of 3
#>  $ el1: int [1:2] 1 2
#>  $ el2: logi [1:2] TRUE FALSE
#>  $ el3:List of 2
#>   ..$ el3_el1: int [1:2] 3 4
#>   ..$ el3_el2: int [1:2] 5 6
# Show the names of the list elements
names(l1)
#> [1] "el1" "el2" "el3"
```

If the list is long, the elements can be named using the names() function, similar to what was carried out for atomic vectors. Here, as a starting point, we take the original unnamed list from the initial example.

```
# Create a list
l2 <- list(1:2,
          c(TRUE, FALSE),
          list(3:4,5:6))
# Name the list elements using names()
names(l2) <- c("el1","el2","el3")
str(l2)
#> List of 3
#>  $ el1: int [1:2] 1 2
#>  $ el2: logi [1:2] TRUE FALSE
#>  $ el3:List of 2
#>   ..$ : int [1:2] 3 4
#>   ..$ : int [1:2] 5 6
```

Note that in this example, the inner list elements for element three are not named; to do this, we would have to access that element directly, and use the `names()` function. Accessing individual list elements is discussed in Chapter 3.

### 2.4.4    Converting a list to an atomic vector

A list can be converted to an atomic vector using the `unlist()` function. This *flattens* the entire list structure, and coercion is performed to ensure that all the elements are of the same type. Below we define the list `l3`, and call `unlist()`, we can explore the output.

```
# Create a list
l3 <- list(1:4,c(TRUE, FALSE),list(2:3,6:7))
# Convert to an atomic vector
l3_av <- unlist(l3)
# Show the result and the type
l3_av
#>  [1] 1 2 3 4 1 0 2 3 6 7
typeof(l3_av)
#> [1] "integer"
```

The result is what you might expect from this process. The logical values are coerced to integer values, all the list values are present, and their order reflects the order in the original list.

---

## 2.5    Mini-case: Two-dice rolls with atomic vectors

The aim of this example is to see how atomic vectors can be used to simulate the rolling of two dice, and to explore whether the expected frequency of outcomes is observed. The following table summarizes the range of outcomes when summing the two dice outcomes, where the total sample space is 36 ($6 \times 6$). In our simulation, we will throw the two dice 10,000 times.

| Dice Rolls | Probability | Sum | Proportion |
|---|---|---|---|
| (1,1) | 1/36 | 2 | 0.02777778 |
| (1,2)(2,1) | 2/36 | 3 | 0.05555556 |
| (1,3)(3,1)(2,2) | 3/36 | 4 | 0.08333333 |
| (1,4)(4,1)(2,3)(3,2) | 4/36 | 5 | 0.1111111 |
| (1,5)(5,1)(2,4)(4,2)(3,3) | 5/36 | 6 | 0.1388889 |
| (1,6)(6,1)(2,5)(5,2)(4,3) (3,4) | 6/36 | 7 | 0.1666667 |
| (2,6)(6,2)(3,5)(5,3)(4,4) | 5/36 | 8 | 0.1388889 |
| (3,6)(6,3)(4,5)(5,4) | 4/36 | 9 | 0.1111111 |

| Dice Rolls | Probability | Sum | Proportion |
|---|---|---|---|
| (4,6)(6,4)(5,5) | 3/36 | 10 | 0.08333333 |
| (3,6)(6,3)(4,5)(5,4) | 2/36 | 11 | 0.05555556 |
| (3,6)(6,3)(4,5)(5,4) | 1/36 | 12 | 0.02777778 |

Here are the steps for the solution. We use the R function `sample()` to generate the dice rolls, and before that, call the function `set.seed()` in order to ensure that the same stream of random numbers are generated. The default behavior of the `sample()` function is that each element has the same chance of being selected. Replacement is set to `TRUE`, which means that once an item is drawn, it can be selected again in the following sample, which makes sense for repeated throws of a die.

```
# set the seed to 100, for replicability
set.seed(100)
# Create a variable for the number of throws
N <- 10000
# generate a sample for dice 1
dice1 <- sample(1:6, N, replace = T)
# generate a sample for dice 2
dice2 <- sample(1:6, N, replace = T)
```

We can observe the first six values from each vector, using the R function `head()`. Note, that the function `tail()` will display the final six values from the vector. These two functions are widely used in R, across a range of data structures.

```
# Information on dice1
head(dice1)
#> [1] 2 6 3 1 2 6
summary(dice1)
#>    Min. 1st Qu.  Median    Mean 3rd Qu.    Max.
#>    1.00    2.00    3.00    3.49    5.00    6.00
# Information on dice2
head(dice2)
#> [1] 4 5 2 5 4 2
summary(dice1)
#>    Min. 1st Qu.  Median    Mean 3rd Qu.    Max.
#>    1.00    2.00    3.00    3.49    5.00    6.00
```

Next, we use the vectorization capability of atomic vectors to sum both vectors, so that the first dice roll for `dice1` is added to corresponding first dice roll for `dice2`, and so on until the final rolls in each vector are summed.

```
# Create a new variable dice_sum, a vectorized sum of both dice.
dice_sum <- dice1 + dice2
head(dice_sum)
#> [1]  6 11  5  6  6  8


summary(dice_sum)
#>    Min. 1st Qu.  Median    Mean 3rd Qu.    Max.
#>    2.00    5.00    7.00    7.01    9.00   12.00
```

Now that we have our final dataset (the simulation of 10,000 rolls for two dice), we can now perform an interesting analysis on the data, using the inbuilt `table()` function in R. This function presents frequencies for each of the values, and is widely used to analyze data in R.

```
# Show the frequencies for the summed values
freq <- table(dice_sum)
freq
#> dice_sum
#>    2    3    4    5    6    7    8    9   10   11   12
#>  274  569  833 1070 1387 1687 1377 1165  807  534  297
```

Finally, the proportion of each sum can be calculated using the vectorization mechanism of R, and also by having flexibility through the use of the function `length()`. These proportions are compared to the expected values, as per statistical theory, and the differences shown, where the values are rounded to five decimal places using R's `round()` function.

```
# Show the frequency proportions for the summed values,
# using the vectorized division operator
sim_probs <- freq/length(dice_sum)
sim_probs
#> dice_sum
#>       2       3       4       5       6       7       8       9      10
#>  0.0274  0.0569  0.0833  0.1070  0.1387  0.1687  0.1377  0.1165  0.0807
#>      11      12
#>  0.0534  0.0297


# Define the exact probabilities for the sum of two dice throws
exact <- c(1/36, 2/36, 3/36, 4/36, 5/36, 6/36,
           5/36, 4/36, 3/36, 2/36, 1/36)


# Use vectorized subtraction to show the differences,
# rounded to 5 decimal places.
round(sim_probs - exact, 5)
```

```
#> dice_sum
#>         2         3         4         5         6         7         8
#> -0.00038   0.00134  -0.00003  -0.00411  -0.00019   0.00203  -0.00119
#>         9        10        11        12
#>   0.00539  -0.00263  -0.00216   0.00192
```

Although it is a short example, there are a number of points worth reflecting on:

- The practical use of vectorization operations as part of the solution. First, two (identically sized) vectors `dice1` and `dice2` were added together, to produce the vector `dice_sum`. Second, the vector `freq`, which was produced by R's `table()` function, was divided by the length of `dice_sum`, and used to calculate the proportions for each outcome.

- The utility of R's `sample()` function, which can generate simulated data that can be helpful for exploring R functions, and for data manipulation.

- The use of the function `set.seed()`, as this allows for exact replication of results by others, and so supports sharing of insights. When you run this example on your own computer, you should get the exact same results.

- When datasets are large, the use of `summary()` provides a summary description of data, and the functions `head()` and `tail()` provide a slice of data at the top, and at the end of the dataset.

## 2.6 Summary of R functions from Chapter 2

A summary of the functions introduced in this chapter now shown. As can be observed, already we are building up a collection of functions that can help us explore and analyze data. This is key to building skills in R, namely, to identify functions that can operate on data structures such as vectors, and include these as part of your analytics workflow. In later chapters, we will explore how you can create your own functions, which can then be used alongside existing R functions. At any point when coding in R, if you wish to view a full description for a function, type in `?<name_of_function>` at the console.

| Function | Description |
|---|---|
| c() | Create an atomic vector. |
| head() | Lists the first six values of a data structure. |
| is.logical() | Checks whether a variable is of type logical. |
| is.integer() | Checks whether a variable is of type integer. |
| is.double() | Checks whether a variable is of type double. |
| is.character() | Checks whether a variable is of type character. |
| is.na() | A function to test for the presence of NA values. |
| ifelse() | Vectorized function that operates on atomic vectors. |
| list() | A function to construct a list. |
| length() | Returns the length of an atomic vector or list. |
| mean() | Calculates the mean for values in a vector. |
| names() | Display or set the vector names. |
| paste0() | Concatenates vectors after converting to a character. |
| str() | Displays the internal structure of a variable. |
| set.seed() | Initializes a pseudorandom number generator. |
| sample() | Generates a random sample of values. |
| summary() | Provides an informative summary of a variable. |
| tail() | Lists the final six values of a data structure. |
| table() | Builds a table of frequency data from a vector. |
| typeof() | Displays the atomic vector type. |
| unlist() | Converts a list to an atomic vector. |

## 2.7   Exercises

The goal of these short exercises is to practice key ideas from the chapter. In most cases, the answers are provided as part of the output; the challenge is to find the right R code that will generate the results.

1.  Predict what the types will be for the following variables, and then verify your results in R.

```
v1 <- c(1L, FALSE)
v2 <- c(1L, 2.0, FALSE)
v3 <- c(2.0, FALSE, "FALSE")
v4 <- c(1:20, seq(1,10,by=.5))
v5 <- unlist(list(1:10,list(11:20,"Hello")))
```

2.  Create the following atomic vector, which is a combination of the character string *Student* and a sequence of numbers from 1 to 10.

Explore how the R function `paste0()` can be used to generate the solution. Type `?paste0` to check out how this function can generate character strings.

```
# The output generated following the call to paste0()
slist
#>  [1] "Student-1"  "Student-2"  "Student-3"  "Student-4"
#>  [5] "Student-5"  "Student-6"  "Student-7"  "Student-8"
#>  [9] "Student-9"  "Student-10"
```

3. Use the R constants (character vectors) LETTERS and letters to generate the following list of four elements, where each list element is a sequence of six alphabetic characters. The names for each list element should be set based on the length of the list.

```
# The new list of four elements
str(l_list)
#> List of 4
#>  $ A: chr [1:6] "a" "b" "c" "d" ...
#>  $ B: chr [1:6] "g" "h" "i" "j" ...
#>  $ C: chr [1:6] "m" "n" "o" "p" ...
#>  $ D: chr [1:6] "s" "t" "u" "v" ...
```

4. Generate a random sample of 20 temperatures (assume integer values in the range −5 to 30) using the `sample()` function (`set.seed(99)`). Assume that temperatures less than 4 are cold, temperatures greater that 25 are hot, and all others are medium; use the `ifelse()` function to generate the following vector. Note that an `ifelse()` call can be nested within another `ifelse()` call.

```
# The temperature dataset
temp
#>  [1] 27 16 29 28 26  7 14 30 25 -2  3 12 18 24 16 14 26  8 -2  8

# The descriptions for each temperature generated by ifelse() call
des
#>  [1] "Hot"     "Medium"  "Hot"     "Hot"     "Hot"     "Medium"
#>  [7] "Medium"  "Hot"     "Medium"  "Cold"    "Cold"    "Medium"
#> [13] "Medium"  "Medium"  "Medium"  "Medium"  "Hot"     "Medium"
#> [19] "Cold"    "Medium"
```

5. Use the expression `set.seed(100)` to ensure that you replicate the result as shown below. Configure a call to the function `sample()` that will generate a sample of 1000 for three categories of people: *Young*, *Adult*, and *Elderly*. Make use of the `prob` argument in `sample()`

(which takes a vector probability weights for obtaining the elements of the vector being sampled) to ensure that 30% *Young*, 40% *Adult* and 30% *Elderly* are sampled. Use the `table()` function to generate the following output (assigned to the variable `ans`). Also, show the proportions for each category.

```
# A summary of the sample (1000 elements),
# based on the probability weights
ans
#> pop
#>   Adult Elderly   Young
#>     399     300     301

# The proportions of each age
prop
#> pop
#>   Adult Elderly   Young
#>   0.399   0.300   0.301
```

# 3

# Subsetting Vectors

R's subsetting operators are fast and powerful. Mastering them allows you to succinctly perform complex operations in a way that few other languages can match.

— Hadley Wickham (Wickham, 2019)

## 3.1 Introduction

Subsetting operators allow you to process data stored in atomic vectors and lists and R provides a range of flexible approaches that can be used to subset data. This chapter presents important subsetting operations, and knowledge of these is key in terms of understanding later chapters on processing lists in R. Upon completing the chapter, you should understand:

- The four main ways to subset a vector, namely, by positive integer, the minus sign, logical vectors, and vector element names.
- The common features, and differences, involved in subsetting lists, and subsetting atomic vectors.
- The role of the [[ operator for processing lists, and how to distinguish this from the [ operator.
- The use of the $ operator, and how it is essentially a convenient shortcut for the [[ operator, and can also be used to add new elements to a list.
- How to use the for loop structure to iterate through a list, in order to process list data. Note that when we introduce *functionals* in later chapters, our reliance on the for loop will significantly reduce.
- How to use the if statement when processing lists, and how that statement differs from the vectorized ifelse() function covered in Chapter 2.

- Additional R functions that allow you to process vectors.
- How to solve all five test exercises.

### Chapter structure

- Section 3.2 describes four ways to subset an atomic vector, and illustrates this using a dataset of customer arrivals at a restaurant, based on random numbers generated from a Poisson distribution.
- Section 3.3 shows how to subset lists, and illustrates an important difference in subsetting operations, namely, how we can return a list, and how we can extract the list contents.
- Section 3.4 introduces a looping statement, and shows how this can be used to process elements within a list. In addition to this, the `if` statement is summarized, as this can be used when processing data.
- Section 3.5 explores a mini-case for subsetting based on a list of Star Wars movies.
- Section 3.6 provides a summary of all the functions introduced in the chapter.
- Section 3.7 provides a number of short coding exercises to test your understanding of the material covered.

## 3.2   Subsetting atomic vectors

In order to demonstrate how to subset from atomic vectors, we will use a business-related problem: exploring a sequence of simulated data that models the daily number of customers arriving at a restaurant, over a 10-day period. Let's assume this arrival data follows a *Poisson distribution*, with a mean ($\lambda$) of 100 customers per day ($\lambda=100$). The Poisson distribution describes a discrete random variable, and has a wide range of applications in operations research, and the mean and variance are both equal to $\lambda$ (Montgomery, 2007).

In R, the `rpois()` function can be used to generate random numbers from a Poisson distribution, with mean ($\lambda$). We call `set.seed()` to ensure the numbers can be replicated, and also name each vector element to be the day number, i.e., "D1", "D2", ...,"D10", using the function `paste0()`. The following code generates our (integer) atomic vector.

```r
# set the seed
set.seed(111)
# Generate the count data, assume a Poisson distribution
customers <- rpois(n = 10, lambda = 100)
# Name each successive element to be the day number
names(customers) <- paste0("D",1:10)
customers
```

```
#>  D1  D2  D3  D4  D5  D6  D7  D8  D9 D10
#> 102  96  97  98 101  85  98 118 102  94
```

For vectors in R, the operator [ is used to subset vectors, and there are a number of ways in which this can be performed (Wickham, 2019).

### 3.2.1  Positive integers

Positive integers will subset atomic vector elements at given locations, and this type of index can have one or more values. To extract the $n^{th}$ item from a vector x the term x[n] is used, and this can also apply to a sequence, for example, a sequence of values, starting at n and finishing at m can be extracted from the vector x as x[n:m]. Indices can also be generated using the combine function c(), which is then used to subset a vector. Any integer vector value(s) can be used to subset a vector.

Here are some examples of how we can subset the vector customers. Note, because we have named the elements in customers, the names appear above each vector element's value. To remove the vector names, the function unname() can be used.

```
# Get the customer from day 1
customers[1]
#>  D1
#> 102


# Get the customers from day 1 through day 5
customers[1:5]
#>  D1  D2  D3  D4  D5
#> 102  96  97  98 101


# Use c() to get the customers from day 1 and the final day
customers[c(1,length(customers))]
#>  D1 D10
#> 102  94


# Note, with c(), any duplicates will be returned
customers[c(1:3,3,3)]
#>  D1  D2  D3  D3  D3
#> 102  96  97  97  97
```

### 3.2.2  Negative integers

Negative integers can be used to exclude elements from a vector, and one or more elements can be excluded. Here are some examples using the customers atomic vector.

```
# Exclude the first day's observation
customers[-1]
#>  D2  D3  D4  D5  D6  D7  D8  D9 D10
#>  96  97  98 101  85  98 118 102  94

# Exclude the first and last day
customers[-c(1,length(customers))]
#>  D2  D3  D4  D5  D6  D7  D8  D9
#>  96  97  98 101  85  98 118 102

# Exclude all values except the first and last day
customers[-(2:(length(customers)-1))]
#>  D1 D10
#> 102  94
```

### 3.2.3   Logical vectors

Logical vectors can be used to subset a vector, and this is a powerful feature
of R, as it allows for the use of relational and logical operators to perform
subsetting. The idea is a simple one: when a logical vector is used to subset a
vector, only the corresponding cells of the target vector that are TRUE in the
logical vector will be retained. Therefore, we can construct a logical vector
based on the use of relational or logical operators on a vector, and then use
this result to subset the vector.

For example, the code below will find any value that is greater than 100.

```
# Create a logical vector based on a relation expression
lv <- customers > 100
lv
#>     D1     D2     D3     D4     D5     D6     D7     D8     D9    D10
#>   TRUE  FALSE  FALSE  FALSE   TRUE  FALSE  FALSE   TRUE   TRUE  FALSE

# Filter the original vector based on the logical vector
customers[lv]
#>  D1  D5  D8  D9
#> 102 101 118 102
```

Typically, these two statements are combined into the one expression, so you
will often see the following style of code in R.

```
# Subset the vector to only show values great than 100
customers[customers > 100]
#>  D1  D5  D8  D9
#> 102 101 118 102
```

A convenient feature of subsetting with logical vectors is that the logical vector size does not have to equal the size of the target vector. When the length of the logical vector is less than the target vector, R will *recycle* the logical vector by repeating the sequence of values until all the target values have been subsetted. For example, this code can be used to extract every second element from a vector.

```
# Subset every second element from the vector
customers[c(TRUE,FALSE)]
#>  D1  D3  D5  D7  D9
#> 102  97 101  98 102
```

## 3.2.4 Named elements

If a vector has named elements, often set via the function `names()`, then elements can subsetted using their names. This is convenient if you need to retrieve an element but do not know its exact indexed location. With the `customers` atomic vector, we have already named each element; therefore, we can use this information to perform subsetting.

```
# Show the vector unnamed
unname(customers)
#>  [1] 102  96  97  98 101  85  98 118 102  94
# Show the vector with the element names
customers
#>  D1  D2  D3  D4  D5  D6  D7  D8  D9 D10
#> 102  96  97  98 101  85  98 118 102  94
# Show the value from day 10
customers["D10"]
#> D10
#>  94
```

In order to return more than one value, we can extend the character vector size by using the `c()` function.

```
# Extract the first and last elements
customers[c("D1","D10")]
#>  D1 D10
#> 102  94
```

In addition to these methods for subsetting an atomic vector, there is another approach that can be used. The R function `which()` returns the TRUE indices of a logical object, and these indices can then be used to extract values.

```
# Use which() to find the indices of the true elements in the
# logical vector
v <- which(customers > 100)
v
```

```
#> D1 D5 D8 D9
#>  1  5  8  9

# Filter customers based on these indices
customers[v]
#>  D1  D5  D8  D9
#> 102 101 118 102
```

## 3.3  Subsetting lists

Subsetting lists is more challenging than subsetting atomic vectors. There are
three methods that can be used, and to illustrate the core idea (before moving
on to a more detailed example), we define a list l1 that contains three elements:
a character vector, an integer vector, and a list, which in turn contains a logical
vector, and a character vector. Each list element is named, for convenience,
and also to illustrate how one of the list subset operators works.

```
# Create a simple vector
l1 <- list(a="Hello",
           b=1:5,
           c=list(d=c(T,T,F),
                  e="Hello World"))
# Show the structure
str(l1)
#> List of 3
#>  $ a: chr "Hello"
#>  $ b: int [1:5] 1 2 3 4 5
#>  $ c:List of 2
#>   ..$ d: logi [1:3] TRUE TRUE FALSE
#>   ..$ e: chr "Hello World"
```

The three subsetting mechanisms for a list are:

- The single square bracket [.
- The double square bracket [[.
- The tag operator $.

When applied to a list, the [ operator will *always return a list*. The same
indexing method used for atomic vectors can also be used for filtering lists,
namely, positive integers, negative integers, logical vectors, and the element
name. Here are examples of filtering a list using each of these methods.

```
# extract the first and third element of the list l1
str(l1[c(1,3)])
```

```
#> List of 2
#>  $ a: chr "Hello"
#>  $ c:List of 2
#>   ..$ d: logi [1:3] TRUE TRUE FALSE
#>   ..$ e: chr "Hello World"

# exclude the first and third element of the list
str(l1[-c(1,3)])
#> List of 1
#>  $ b: int [1:5] 1 2 3 4 5

# Show the first and third element using a logical vector
str(l1[c(TRUE,FALSE,TRUE)])
#> List of 2
#>  $ a: chr "Hello"
#>  $ c:List of 2
#>   ..$ d: logi [1:3] TRUE TRUE FALSE
#>   ..$ e: chr "Hello World"

# Show the first element using a character vector
str(l1["a"])
#> List of 1
#>  $ a: chr "Hello"
```

Note that all examples with [ return a list, but in many cases this is not sufficient for analysis, as we will need to access the data within the list (which can be an atomic vector, and also could be another list). For example, finding the value of element a or element b. To do this, we must use the [[ operator, which extracts the *contents of a list* at a given location (i.e., element 1, 2, .., N), where N is the list length. The following short examples show how we can access the data within a list.

```
# extract the contents of the first list element
l1[[1]]
#> [1] "Hello"

# extract the contents of the second  list element
l1[[2]]
#> [1] 1 2 3 4 5

# extract the contents of the third list element (a list!)
str(l1[[3]])
#> List of 2
#>  $ d: logi [1:3] TRUE TRUE FALSE
#>  $ e: chr "Hello World"
```

```
# extract the contents of the first element of the third element
l1[[3]][[1]]
#> [1]  TRUE  TRUE FALSE
```

```
# extract the contents of the second element of the third element
l1[[3]][[2]]
#> [1] "Hello World"
```

The list element names can also be used to subset the contents, for example:

```
# extract the contents of the first list element
l1[["a"]]
#> [1] "Hello"
```

```
# extract the contents of the second  list element
l1[["b"]]
#> [1] 1 2 3 4 5
```

```
# extract the contents of the third list element (a list!)
str(l1[["c"]])
#> List of 2
#>  $ d: logi [1:3] TRUE TRUE FALSE
#>  $ e: chr "Hello World"
```

```
# extract the contents of the first element of the third element
l1[["c"]][["d"]]
#> [1]  TRUE  TRUE FALSE
```

```
# extract the contents of the second element of the third element
l1[["c"]][["e"]]
#> [1] "Hello World"
```

There is a convenient alternative to the [[ operator, and this is the tag operator $ which can be used once a list element is named. For example, for our list l1 the terms l1[[1]], l1[["a"] and l1$a return the same result. In the general case my_list[["y"]] is equivalent to my_list$y. Using the $ operator also is convenient as it re-appears later when we discuss another R data structure, data frames, in Chapter 5. Here are examples of how to use $ based on the previous example. Notice that the exact same results are returned.

```
# extract the contents of the first list element
l1$a
#> [1] "Hello"
```

```
# extract the contents of the second  list element
l1$b
```

```
#> [1] 1 2 3 4 5

# extract the contents of the third list element (a list!)
str(l1$c)
#> List of 2
#>  $ d: logi [1:3] TRUE TRUE FALSE
#>  $ e: chr "Hello World"

# extract the contents of the first element of the third element
l1$c$d
#> [1]  TRUE  TRUE FALSE

# extract the contents of the second element of the third element
l1$c$e
#> [1] "Hello World"
```

We can now move on to explore a more realistic example, where the variable products is a product list that stores product sales information.

```
# A small products database. Main list has two products
products <- list(A=list(product="A",
                        sales=12000,
                        quarterly=list(quarter=1:4,
                                       sales=c(6000,3000,2000,1000))),
                 B=list(product="B",
                        sales=8000,
                        quarterly=list(quarter=1:4,
                                       sales=c(2500,1500,2800,1200))))
str(products)
#> List of 2
#>  $ A:List of 3
#>   ..$ product  : chr "A"
#>   ..$ sales    : num 12000
#>   ..$ quarterly:List of 2
#>   .. ..$ quarter: int [1:4] 1 2 3 4
#>   .. ..$ sales  : num [1:4] 6000 3000 2000 1000
#>  $ B:List of 3
#>   ..$ product  : chr "B"
#>   ..$ sales    : num 8000
#>   ..$ quarterly:List of 2
#>   .. ..$ quarter: int [1:4] 1 2 3 4
#>   .. ..$ sales  : num [1:4] 2500 1500 2800 1200
```

The list structure is visualized in Figure 3.1, and from this we can make a number of observations:

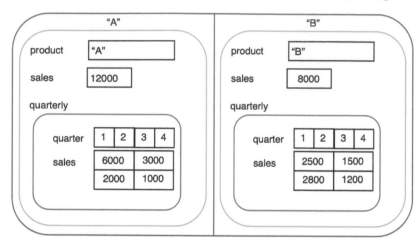

**FIGURE 3.1** Visualization of the product's data structure

- At its core, the list is simply a vector of two named elements, and this is shown with the red lines. We can verify this with R code to show (1) the length of the vector and (2) the name of each element.

```
# Show the vector length (2 elements)
length(products)
#> [1] 2
# Show the names of each element
names(products)
#> [1] "A" "B"
```

- However, even though there are just two elements in the list, the list has significant internal structure. Each element contains a list, highlighted in green. This inner list contains information about an individual product, and it comprises three elements:

  - The product name (product), a character atomic vector.
  - The annual sales amount (sales), a numeric atomic vector.
  - The quarterly sales (quarterly), which is another list (colored blue). This list contains two atomic vector elements: the quarter number (quarter) with values 1 to 4, the corresponding sales amount (sales) for each quarter.

Based on this list structure, we now identify eight different subset examples for the list, which show how we can access the list data in a number of different ways. Before exploring the code, we visualize the different subsetting outputs in Figure 3.2.

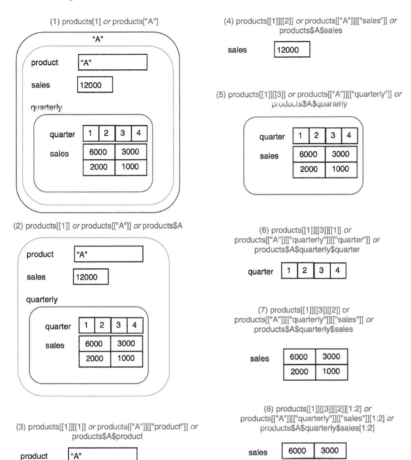

**FIGURE 3.2** Subsetting lists

The eight examples are:

1. Extract the first element of the list as a *list*, using [ in two ways, one with a *positive integer* to retrieve the element, the other using the *element name* "A". Each of these subsetting instructions returns a list structure, and the list size is equal to 1.

```
# Example (1) - get the first element of the list as a list
ex1.1 <- products[1]
ex1.2 <- products["A"]
str(ex1.1)
#> List of 1
#>  $ A:List of 3
#>   ..$ product  : chr "A"
```

```
#>    ..$ sales    : num 12000
#>    ..$ quarterly:List of 2
#>    .. ..$ quarter: int [1:4] 1 2 3 4
#>    .. ..$ sales  : num [1:4] 6000 3000 2000 1000
```

2. Extract the contents of the first list element. Use [[ in two ways, one with a *positive integer* to retrieve the element, the other using the *element name* "A". Each of these subsetting commands returns a list of size 3, where the three elements are: the product (a character vector), the sales (numeric vector), and a list of two elements, which contain two vectors, one for the quarter, the other for the sales in each quarter. We also show the tag operator to extract the same information, namely products$A.

```
# Example (2) - get the contents of the first list element
ex2.1 <- products[[1]]
ex2.2 <- products[["A"]]
ex2.3 <- products$A
str(ex2.1)
#> List of 3
#>  $ product  : chr "A"
#>  $ sales    : num 12000
#>  $ quarterly:List of 2
#>    ..$ quarter: int [1:4] 1 2 3 4
#>    ..$ sales  : num [1:4] 6000 3000 2000 1000
```

3. Extract the product name from the first list element (which is a list), and this requires the use of two [[ operators, the first of these extracts the first element of products, and the second extracts the first element of this list. Again, there are three ways to access this information: the index location, the vector names, or the use of the $ operator. The value returned is a character vector.

```
# Example (3) - get the product name for the first product
ex3.1 <- products[[1]][[1]]
ex3.2 <- products[["A"]][["product"]]
ex3.3 <- products$A$product
str(ex3.1)
#>  chr "A"
```

4. In a similar way, we can also gather the annual sales of product A, by extracting the second element of the inner list, and again, use three different mechanisms to access the same information.

```
# Example (4) - get the annual sales for the first product
ex4.1 <- products[[1]][[2]]
ex4.2 <- products[["A"]][["sales"]]
ex4.3 <- products$A$sales
str(ex4.1)
#> num 12000
```

5. The third element of the first element is also a list, and when we extract this we see that the structure contains two atomic vectors.

```
# Example (5) - get as a list, the detailed quarterly sales
ex5.1 <- products[[1]][[3]]
ex5.2 <- products[["A"]][["quarterly"]]
ex5.3 <- products$A$quarterly
str(ex5.1)
#> List of 2
#>  $ quarter: int [1:4] 1 2 3 4
#>  $ sales  : num [1:4] 6000 3000 2000 1000
```

6. For the list extracted in the previous step, we can then access the inner list values, again by using the [[ or $ operators. Notice that we now use three successive [[ calls, because we have three lists that can be accessed. For this example, the numeric vector that contains the quarters is returned.

```
# Example (6) - get the quarters
ex6.1 <- products[[1]][[3]][[1]]
ex6.1 <- products[["A"]][["quarterly"]][["quarter"]]
ex6.1 <- products$A$quarterly$quarter
str(ex6.1)
#>  int [1:4] 1 2 3 4
```

7. In a similar way, we can access the quarterly sales, as they are contained in the second vector within the list.

```
# Example (7) - get the quarterly sales
ex7.1 <- products[[1]][[3]][[2]]
ex7.1 <- products[["A"]][["quarterly"]][["sales"]]
ex7.1 <- products$A$quarterly$sales
str(ex7.1)
#>  num [1:4] 6000 3000 2000 1000
```

8. Finally, we can then use conventional vector notation to subset the quarterly sales values. While the assignment statement for ex8.1

is not very user-friendly, and is also dependent on knowing the exact index location for each element, it's a valuable exercise to test your knowledge of how to subset a detailed list structure. These commands extract the quarterly sales for quarters one and two.

```
# Example (8) - get the quarterly sales for the first two quarters
ex8.1 <- products[[1]][[3]][[2]][1:2]
ex8.2 <-products[["A"]][["quarterly"]][["sales"]][1:2]
ex8.3 <-products$A$quarterly$sales[1:2]
str(ex8.1)
#>  num [1:2] 6000 3000
```

An additional point that is worth making is: an interesting feature of the $ operator is that it can be used to add a new element to the list, or overwrite an existing element. For example, we can overwrite and also add extra information to product A.

```
# Increase the sales of product A by 10,000
products$A$sales <- products$A$sales + 10000
# Add a new field to product A
products$A$type <- "Food"
str(products$A)
#> List of 4
#>  $ product  : chr "A"
#>  $ sales    : num 22000
#>  $ quarterly:List of 2
#>   ..$ quarter: int [1:4] 1 2 3 4
#>   ..$ sales  : num [1:4] 6000 3000 2000 1000
#>  $ type     : chr "Food"
```

So overall, some observations on list subsetting are as follows:

- Clearly, for list manipulation, the tag operator is the most programmer-friendly, so it is recommended to use this and also try to ensure that the list elements are named.
- Indexing by positive integer is convenient for looping structures; we will see an example of this shortly.
- In Chapter 10 we will introduce flexible functions such as those in the package purrr that provide efficient and impressive ways to iterate through lists.

Before that, we will show how to iterate through a list from *first principles* using a loop structure, along with the index position of the list.

## 3.4   Iteration using loops, and the `if` statement

Iteration is a core structure in all programming languages, and R is no exception. There are a number of basic looping structures than can be used in R, and we will focus on one of these, the `for` loop. The general structure is `for(var in seq)expr`, where:

- `var` is a name for a variable that will change its value for each loop iteration.
- `seq` is an expression that evaluates to a vector.
- `expr` is an expression, which can be either a simple expression, or a compound expression of the form `{expr1; expr2}`, which is effectively a number of lines of code with two curly braces.

A convenient method to iterate over a vector (a list or an atomic vector), is to use the function `seq_along()` which returns the indices of a vector. For example, consider the vector v below, which contains a simulation of ten dice rolls.

```
set.seed(100)
v <- sample(1:6,10,replace = T)
v
#>  [1] 2 6 3 1 2 6 4 6 6 4

sa <- seq_along(v)
sa
#>  [1]  1  2  3  4  5  6  7  8  9 10
```

Notice that the vector sa returns the set of indices of v, starting a 1 and finishing at 10. This is helpful, because we can use this result to iterate through every element of v and perform a calculation. For example, let's find the number of elements in v that equal six, and do this using a loop. (Of course we could just type `sum(v==6)` and that would give us the same answer.) In the loop shown below, the value of i takes on the current value in the sequence, starting at 1 and finishing at 10.

```
n_six <- 0
for(i in seq_along(v)){
  n_six <- n_six + as.integer(v[i] == 6)
}
n_six
#> [1] 4
```

The looping structure is valuable when we are dealing with lists, because we can use the `[[` operator, along with the loop index, to extract values. So let's try the following example by revisiting to the products list, which is a list of

two elements. Our goal is to find the average sales for the two products, and
for this we can use a list.

```
# Initialize the total to be 0
sum_sales <- 0
# Iterate through the products list (2 elements) using seq_along()
for(i in seq_along(products)){
  # Increment the number of sales by the current product sales value
  sum_sales <- sum_sales+products[[i]]$sales
}
# Calculate the average
avr_sales <- sum_sales / length(products)
# Display the average
avr_sales
#> [1] 15000
```

When iterating through individual vector elements, you may need to execute a
statement based on a current vector value. To do this, the if statement can
be used, and there are two main forms:

- if(cond) expr which evaluates expr if the condition cond is true
- if(cond) true.expr else false.expr, which evaluates true.expr if the condition is true, and otherwise evaluates false.expr

If the variable used in the conditional expression has a length greater than 1,
only the first element will be used.

Here are some examples, where we can iterate through a vector to find those
values that are greater than the mean. Note that the vectorized ifelse()
function would normally be used for this type of processing.

```
# create a test vector
v <- 1:10
# create a logical vector to store the result
lv <- vector(mode="logical",length(v))
# Loop through the vector, examining each element
for(i in seq_along(v)){
  if(v[i] > mean(v))
    lv[i] <- TRUE
  else
    lv[i] <- FALSE
}
v[lv]
#> [1]  6  7  8  9 10

# The following code would have the same result
# using the vectorized ifelse function
lv1 <- ifelse(v>mean(v),T,F)
```

```
v[lv1]
#> [1]  6  7  8  9 10

# This would also generate the same result
v[v>mean(v)]
#> [1]  6  7  8  9 10
```

In common with the `for` loop, curly braces can be used if there is more than one statement to be executed. The following mini-case will also show an example of iterating through a list using the `for` statement.

---

## 3.5   Mini-case: Star Wars movies

The aim of this mini-case is to show how we can subset lists. The CRAN package `repurrrsive` is used (and you must install this package yourself), and this includes lists of R objects for use in teaching and examples, and it contains entities from the Star Wars universe, for example, lists on films (`sw_films`), people (`sw_people`), planets (`sw_planets`), species (`sw_species`), and starships (`sw_starships`). Here we focus on the `sw_films` list, which contains seven elements, and each element is a list that contains 14 elements.

For example, the first element in the list is shown below (movie title "A New Hope"), and as can be seen, most of the elements are character strings, some of which contains web links to the Star Wars application programmer interface SWAPI (calling this is outside of the scope of this mini-case).

```
library(repurrrsive)
length(sw_films)
#> [1] 7
# show the list elements for 1st element
names(sw_films[[1]])
#>  [1] "title"      "episode_id"   "opening_crawl" "director"
#>  [5] "producer"   "release_date" "characters"    "planets"
#>  [9] "starships"  "vehicles"     "species"       "created"
#> [13] "edited"     "url"
```

Given that the list `sw_films` is a *list of lists*, we can access the data by first accessing the contents of each list element (which returns a list), and then accessing the contents of that list, which then contains 14 different atomic vectors. For example, to retrieve the movie name and film director of the first and last movies, the following commands can be used.

```
# Get the first film name and movie director
sw_films[[1]][[1]]
```

```
#> [1] "A New Hope"
sw_films[[1]][[4]]
#> [1] "George Lucas"

# Get the last film name and movie director
sw_films[[length(sw_films)]][[1]]
#> [1] "The Force Awakens"
sw_films[[length(sw_films)]][[4]]
#> [1] "J. J. Abrams"
```

However, given that the inner list is defined with elements names (which is convenient and means that we don't have to know the position of each element), the following code will also retrieve the same data.

```
# Get the first film name and movie director
sw_films[[1]]$title
#> [1] "A New Hope"
sw_films[[1]]$director
#> [1] "George Lucas"

# Get the last film name and movie director
sw_films[[length(sw_films)]]$title
#> [1] "The Force Awakens"
sw_films[[length(sw_films)]]$director
#> [1] "J. J. Abrams"
```

Now we will show how to filter the list `sw_films` in order to narrow our search, for example, to find all movies directed by George Lucas. The strategy used is:

- A for-loop structure (along with `seq_along()`) is used to iterate over the entire loop and mark those elements as either a match (TRUE) or not a match (FALSE). This information is stored in an atomic vector. Note that while we use a for-loop structure, in subsequent chapters we will use what are known as *functionals* to iterate over data structures; therefore, after this chapter, we will not see too many more for-loops being used.

- Before entering the loop, we create a logical vector variable (`is_target`) of size 7 (the same size as the list), and this will store information on whether a list item should be marked for further processing.

- For each list element we extract the director's name, check if it matches the target ("George Lucas"), and store this logical value in the corresponding element of `is_target`.

- The vector `is_target` can then be used to filter the original `sw_films` list and retain all the movies directed by George Lucas.

The code to perform this list filtering is shown below.

```
# Search for movies by George Lucas and store these in a new list
target <- "George Lucas"
# Create a logical vector to hold information for positive matches
is_target <- vector(mode="logical",length = length(sw_films))
# Iterate through the entire sw_films list (of 7)
for(i in seq_along(sw_films)){
  is_target[i] <- sw_films[[i]]$director == target
}
target_list <- sw_films[is_target]
```

The results from the logical vector can be viewed, and they show that the first four films in the list match. We also confirm the length of `target_list`.

```
is_target
#> [1]  TRUE  TRUE  TRUE  TRUE FALSE FALSE FALSE
length(target_list)
#> [1] 4
```

Given that we now have a new filtered list, we can proceed to extract information from this list. In this case, we look to extract the movie titles into a new data structure, in this case an atomic vector. The steps here are:

- Create an output data structure which is the length of the target list (in this case four elements)
- Iterate, again using a for-loop, through the list and copy the title into the new list

This code is shown below. We make use of the function `vector()` to create an initial list structure, and this is good practice when you are using a loop to iterate through a list, as we will always know the length of the output, and specifying the data element size up-front is more efficient.

```
# Create a movies vector to store the movie names
movies <- vector(mode="character",length = length(target_list))
# Iterate through the list to extract the movie title
for(i in seq_along(target_list)){
  movies[i]<-target_list[[i]]$title
}
movies
#> [1] "A New Hope"          "Attack of the Clones"
#> [3] "The Phantom Menace"  "Revenge of the Sith"
```

One aspect of R you will learn to appreciate is that there are often many ways to achieve the same outcome. For example, another way to access the movies of George Lucas would be to rearrange the *list of lists* into a single list, where each list element is an atomic vector of values (each of size 7). The process for creating the new data structure is:

- Create a new list (`sw_films1`) of elements you wish to store (for example, movie title, episode_id and director) from the original list. This new list initially contains empty vectors.
- Loop through the `sw_films` list and append each movie title and director to the corresponding element of `sw_films1`

```
# Create a new list to store the data in a different way
sw_films1 <- list(title=c(),
                  episode_id=c(),
                  director=c())
# Iterate through the list to append the title and director
for(i in seq_along(sw_films)){
  sw_films1$title      <- c(sw_films1$title,
                            sw_films[[i]]$title)
  sw_films1$episode_id <- c(sw_films1$episode_id,
                            sw_films[[i]]$episode_id)
  sw_films1$director   <- c(sw_films1$director,
                            sw_films[[i]]$director)
}
sw_films1
#> $title
#> [1] "A New Hope"              "Attack of the Clones"
#> [3] "The Phantom Menace"      "Revenge of the Sith"
#> [5] "Return of the Jedi"      "The Empire Strikes Back"
#> [7] "The Force Awakens"
#>
#> $episode_id
#> [1] 4 2 1 3 6 5 7
#>
#> $director
#> [1] "George Lucas"    "George Lucas"      "George Lucas"
#> [4] "George Lucas"    "Richard Marquand"  "Irvin Kershner"
#> [7] "J. J. Abrams"
```

Notice that we now have one list, and this list has three elements, each an atomic vector of size 7. These vectors can also be viewed as *parallel vectors*, where each vector is the same size, and the i-th elements of each vector are related. For example, location 1 for each vector contains information on "A New Hope".

```
# Showing how parallel vectors work
sw_films1$title[1]
#> [1] "A New Hope"
sw_films1$episode_id[1]
#> [1] 4
```

```
sw_films1$director[1]
#> [1] "George Lucas"
```

This feature can be then exploited to filter related atomic vectors using logical vector subsetting.

```
# Find all the movie titles by George Lucas
sw_films1$title[sw_films1$director=="George Lucas"]
#> [1] "A New Hope"          "Attack of the Clones"
#> [3] "The Phantom Menace"  "Revenge of the Sith"
```

In general, in the remaining book chapters, we will find easier ways to process lists using functionals. The aim here is to show how you can process lists using your own loop structures, along with the function `seq_along()`.

## 3.6 Summary of R functions from Chapter 3

A summary of the functions introduced in this chapter is now shown. All of these functions are part of base R.

| Function | Description |
| --- | --- |
| as.list() | Coerces the input argument into a list. |
| paste0() | Converts and concatenates arguments to character strings. |
| rpois() | Generates Poisson random numbers (mean lambda). |
| seq_along() | Generates a sequence to iterate over vectors. |
| which() | Provide the TRUE indices of a logical object. |

## 3.7 Exercises

These exercises mostly focus on processing lists, given that they are a more complex type of vector. For many of the exercises, a loop structure will be required. As mentioned, the need for using loops will mostly disappear after this chapter, as we will soon introduce a different means of iterating, which is known as a *functional*.

1. In R, a data frame (which will be introduced in later chapters) can be converted to a list, using the function `as.list()`. There is a data frame in R known as `mtcars`, and we fill first convert this to a list, and explore its structure.

```
cars <- as.list(mtcars)
# Display list element names
names(cars)
#>  [1] "mpg"  "cyl"  "disp" "hp"   "drat" "wt"   "qsec" "vs"   "am"
#> [10] "gear" "carb"
# Show length of list
length(cars)
#> [1] 11
# Display sample element (final one)
cars["carb"]
#> $carb
#>  [1] 4 4 1 1 2 1 4 2 2 4 4 3 3 3 4 4 4 1 2 1 1 2 2 4 2 1 2 2 4 6 8 2
```

Notice that this is a list of eleven elements, where each element is a feature
of a specific car. For example mpg represents the miles per gallon for each car,
and disp stores the engine size, or displacement. The list contents are eleven
numeric vectors, each of size 32, and these can be viewed as *parallel arrays*,
where all data in the 1st of each vector refers to car number one, and all data
in final location refers to the final car.

The aim of the exercise is to generate a list to contain the mean value for mpg
and disp. Use the following code to create the list structure.

```
result <- list(mean_mpg=vector(mode="numeric", length=1),
              mean_disp=vector(mode="numeric", length=1))
str(result)
#> List of 2
#>  $ mean_mpg : num 0
#>  $ mean_disp: num 0
```

The following shows the calculated result. Note that the cars list should first
be subsetted to only include those two elements that are required (i.e., mpg
and disp.)

```
str(result)
#> List of 2
#>  $ mean_mpg : num 20.1
#>  $ mean_disp: num 231
```

You can confirm your results with the following commands in R.

```
mean(cars$mpg)
#> [1] 20.09
mean(cars$disp)
#> [1] 230.7
```

2. Filter the list sw_people (87 elements), contained in repurrrsive
   to include only those whose height is *not unknown*. Use an atomic

vector `has_height` to filter the list, and populate this vector using a loop structure. This new list (`sw_people1`) should have 81 elements.

```
sum(has_height)
#> [1] 81
length(sw_people1)
#> [1] 81
```

3. Using a `for` loop over the filtered list `sw_people1` from exercise 2, create a list of people whose height is greater than or equal to 225 inches. The resulting vector should grow as matches are found, as we do not know in advance how many people will be contained in the result. Use the command `characters <- c()` to create the initial empty result vector.

The following result should be obtained.

```
# These are the characters whose height is >= 225
characters
#> [1] "Chewbacca"   "Yarael Poof" "Lama Su"     "Tarfful"
```

4. Using a `for` loop to iterate over the list `sw_planets` and display those planets (in a character vector) whose diameter is greater than or equal to the mean. Use a pre-processing step that will add a new list element to each planet, called `diameter_numeric`. This pre-processing step can also be used to calculate the mean diameter. Also, use the pre-processing step to keep track of those planets whose diameter is "unknown", and use this information to create an updated list `sw_planets1` that excludes all the values.

You can check your solution against the following values.

```
# The list elements that will be excluded (diameter unknown)
exclude
#>  [1] 37 39 42 44 45 46 47 49 50 51 52 53 54 55 57 59 61
# The mean diameter
mean_diameter
#> [1] 8936
# The first three and last three planets returned
gte_mean[c(1:3,(length(gte_mean)-2):(length(gte_mean)))]
#> [1] "Alderaan"   "Yavin IV"   "Bespin"     "Muunilinst" "Kalee"
#> [6] "Tatooine"
```

5. Based on the list `sw_species`, and given that each species has a classification, create the following tabular summary, again using

a loop to iterate through the list. Make use of the `table()` function
that was covered in Chapter 2 to present the summary.

```
# A tabular summary of the types of species
t_species
#> c_species
#>  amphibian artificial  gastropod  insectoid      mammal     mammals
#>          6          1          1          1          16           1
#>    reptile  reptilian   sentient    unknown
#>          3          1          1          6
```

# 4

---

*Functions, Functionals, and the R Pipe*

---

An important advantage of R and other interactive languages is to make programming in the small an easy task. You should exploit this by creating functions habitually, as a natural reaction to any idea that seems likely to come up more than once.

— John Chambers (Chambers, 2017)

---

## 4.1 Introduction

Creating functions is essential to achieving a high productivity return from your use of R, and therefore it's worth spending time on the material in this chapter, as it is foundational for the rest of the book. Functions are building blocks in R, and are small units that can take an input, process it, and return a result. We already have used R functions in previous chapters, for example `sample()`, and these functions all represent a unit of code we can call with arguments, and receive a result.

Upon completing the chapter, you should understand:

- How to write your own function, call it via arguments, and return values based on the last evaluated expression.
- The different ways to pass arguments to functions.
- How to add robustness checking to your functions to minimize the chance of processing errors.
- What an environment is, and the environment hierarchy within R.
- How a function contains a reference to its enclosing environment, and how this mechanism allows a function to access variables in the global environment.

- What a functional is, namely, a function that can take a function as an argument.
- The functional `lapply()`, and how this can be used as an alternative to loops, to iterate over lists and vectors.
- R's native pipe operator, and how this can streamline a sequence of data processing operations.
- How to solve all five test exercises.

### Chapter structure

- Section 4.2 shows how to create a function that accepts inputs, and returns a result.
- Section 4.3 highlights the flexibility of R in the way in which arguments can be passed to functions.
- Section 4.5 provides an insight into the key idea of an environment in R, how R finds objects, and why functions can access variables in the global environment.
- Section 4.4 shows how checks can be made at the start of a function to ensure it is robust, and then exit if the input is not as expected.
- Section 4.6 shows how functions are objects in their own right, and so can be passed to functions. This can be used to iterate over data structures, and replace the for loop, and this approach is illustrated using the function `lapply()`.
- Section 4.7 returns to the mini-case from Chapter 3, and solves the problems using `lapply()` instead of loops.
- Section 4.8 introduces R's pipe operator, which provides a valuable mechanism to chain operations together in a data pipeline in order to improve code clarity.
- Section 4.9 provides a summary of all the functions introduced in the chapter.
- Section 4.10 provides a number of short coding exercises to test your understanding of functions, functionals, environments, and the pipe operator.

## 4.2   Functions

Functions are a fundamental building block of R. A function can be defined as a group of instructions that: takes input, uses the input to compute other values, and returns a result (Matloff, 2011). It is recommended that users of R should adopt the habit of creating simple functions which will make their work more effective and also more trustworthy (Chambers, 2008). Functions are declared using the `function` reserved word, and are objects, which means they can also be passed as arguments to other functions. The general form of a function in R is:

```
function (arguments) expression
```

where:

- `arguments` provides the arguments (inputs) to a function, and are separated by commas,
- `expression` is any legal R expression contained within the function body, and is usually enclosed in curly braces (when there is more than one expression),
- the last evaluated expression is returned by the function, although the function `return()` can be also used to return values.

## 4.2.1   A first function: Returning even numbers

Our first function will take in a vector of numbers, and return only those that are even. To do this, R's modulus operator %% is used, as this returns the remainder of two numbers, following their division. The following snippet shows the results of using the modulus operator, which divides the vector 1:5 by 2 and calculates the remainder.

```
v <- 1:5
x <- v %% 2
x
#> [1] 1 0 1 0 1
```

Before writing the function, we can explore the logic needed to filter the input vector. In this example, when the modulus function returns a remainder of 0, then we know the number is divisible by 2. We can use this information to create a logical vector that can then be used to filter the modulus result, as shown in the following code.

```
# The logical vector where even values are TRUE
lv <- x == 0

# Show the logical vector
lv
#> [1] FALSE  TRUE FALSE  TRUE FALSE

# Filter the original vector
v[lv]
#> [1] 2 4
```

This logic can now be embedded within an R function which we will call `evens()`, which takes in one argument (the original vector), and returns a filtered version of the vector that only includes even numbers. We will take a parsimonious approach to code writing and just limit the function to one line of code. The last evaluated expression (which in this case is the first expression) is returned.

```
evens <- function(v){
  v[v%%2==0]
}


x1 <- 1:7

evens(x1)
#> [1] 2 4 6
```

However, there is no error checking on the input, which is an issue we will provide a solution to in Section 4.4.

## 4.2.2   A second function: Removing duplicates

Our second function takes an approach of building on the work of R developers, and it uses an existing base R function to create a new one. This function will take in a vector of random numbers, and remove any duplicates. To remove the duplicates, we will make use of the R function duplicated(), which returns a logical vector that contains TRUE if a value is duplicated. Here is an example of its use, and we can observe that location 5 is TRUE because this is where the value 2 is duplicated for the first time.

```
set.seed(100)
v <- sample(1:6,10,replace = T)
v
#>   [1] 2 6 3 1 2 6 4 6 6 4
duplicated(v)
#>   [1] FALSE FALSE FALSE FALSE  TRUE  TRUE FALSE  TRUE  TRUE  TRUE
```

In order to find the set of values that are unique, we can use the information returned by duplicated(), as follows.

```
v[!duplicated(v)]
#> [1] 2 6 3 1 4
```

The challenge now is to embed this logic into a function. We will call the function my_unique() that takes in a vector (one argument), and returns the unique values from the vector. It is also useful to write the function into a source file; let's assume the file is called my_functions.R. This function could just be written in one line of code, but we will break it down into a number of separate steps just to clarify the process.

```
my_unique <- function(x){
  # Use duplicated() to create a logical vector
  dup_logi <- duplicated(x)
  # Invert the logical vector so that unique values are set to TRUE
  unique_logi <- !dup_logi
```

```
# Subset x to store those values are unique
ans <- x[unique_logi]
# Evaluate the variable ans so that it is returned
ans
}
```

To load the function into R, call the source function (this is easy to do within RStudio by clicking the "Source" button). The function is then loaded into the workspace, and its existence can be confirmed by calling the `ls()` function, which returns a vector of character strings giving the names of the objects in the specified environment.

```
# The call to source loads the function into the global environment
source("my_functions.R")
```

Once a function is loaded in the environment (we will discuss environments in more detail soon), it can then be accessed by a function call. The code below shows the call, and the call's result is stored in the variable `ans`.

```
set.seed(100)
v <- sample(1:6,10,replace = T)
ans <- my_unique(v)
ans
#> [1] 2 6 3 1 4
```

Normally, when writing in R, that following shorter function would suffice for `my_unique()`.

```
my_unique <- function(x){
  x[!duplicated(x)]
}
```

Note that there is already a function in R to perform this operation, and, as you might expect, this is named `unique()`.

```
unique(v)
#> [1] 2 6 3 1 4
```

Interestingly in R, functions are objects, so they can be passed to functions as arguments, and this feature will be used many times in this textbook, particularly when we consider functions that are known as *functionals*, namely, *functions that accept functions as arguments*. To send a function as an argument, all that is required is the function name. Let's take a simple example of a function called `my_summmary()`, which can take a vector, and a function (e.g., `min()` or `max()`), and then return the result of whatever function is sent in. Inside the function, the variable `fn` contains the reference to the function provided, so for this example, the first time it is called, `fn(v)` is equivalent to `min(v)`.

```
my_summary <- function(v, fn){
  fn(v)
}


# Call my_summary() to get the minimum value
my_summary(1:10,min)
#> [1] 1
# Call my_summary() to get the maximum value
my_summary(1:10,max)
#> [1] 10

my_min<- function(v){
  min(v)
}
```

You can write your own functions that can be passed into another function.

```
# Call my_summary() to get the minimum value
my_summary(1:10,my_min)
#> [1] 1
```

Furthermore, you could also write the logic of my_min as an *anonymous function* (i.e., it is not assigned to a variable, and so does not appear in the global environment), and examples are shown below. Right now this might seem like an odd thing to do; however, it is a key idea used when we start to explore functionals such as lapply(), and later purrrr::map(), to iterate over list structures, and apply an action to each list element.

```
# Call my_summary() using an anonymous function
my_summary(1:10,function(y)min(y))
#> [1] 1
# Call my_summary() using an anonymous function
my_summary(1:10,function(y)max(y))
#> [1] 10
```

## 4.3   Passing arguments to functions

When programming in R, it is useful to distinguish between the *formal arguments*, which are the property of the function itself, and the actual arguments, which can vary when the function is called (Wickham, 2019). For example, the function sum() could be called with different arguments, as shown below.

```
v <- c(1,2,3,NA)
sum(v)
```

```
#> [1] NA
sum(v,na.rm=TRUE)
#> [1] 6
```

Each function in R is defined with a set of formal arguments that have a fixed positional order, and often that is the way arguments are then passed into functions (e.g., by position). However, arguments can also be passed in by *complete name* or *partial name*, and arguments can have default values. We can explore this via the following example for the function f, which has three formal arguments: abc, bcd, and bce, and simply returns an atomic vector showing the function inputs (argument one, argument two, and argument three).

```
f <- function(abc,bcd,bce){
  c(FirstArg=abc,SecondArg=bcd,ThirdArg=bce)
}
```

The three ways to send arguments into the function are:

- *By position*, where the arguments are copied to the corresponding argument location, which is a common method used in programming languages. Here 1 is copied to abc, 2 is copied to bcd, and 3 is copied to bce.

```
f(1,2,3)
#>  FirstArg SecondArg  ThirdArg
#>         1         2         3
```

- *By complete name*, where arguments are first copied to their corresponding name, before other arguments are then copied via their positions. The advantage of this is that the programmer calling the function does not need to know the exact position of an argument to call it, and we have seen this already with the use of the argument na.rm in R base functions such as sum().

```
f(2,3,abc=1)
#>  FirstArg SecondArg  ThirdArg
#>         1         2         3
```

- *By partial name*, where argument names are matched, and where a unique match is found, that argument will be selected. A observation here is that if there is more than one match, the function call will fail. Furthermore, using partial matching can lead to confusion for someone trying to understand the code, so it's probably not a great idea to make use of this argument passing mechanism in your own code.

```
f(2,a=1,3)
#>  FirstArg SecondArg  ThirdArg
#>         1         2         3
```

An important feature with defining arguments is that they can be allocated default values, which provides flexibility in that not all the arguments need to be called each time the function is invoked. We can modify the function f so that each argument has an arbitrary default value.

```
f <- function(abc=1,bcd=2,bce=3){
  c(FirstArg=abc,SecondArg=bcd,ThirdArg=bce)
}
```

Following this, the function can be called in four different ways: with no arguments, and with one, two, or three arguments. In this example, a mixture of positional and complete naming matching are used.

```
f()
#>  FirstArg SecondArg  ThirdArg
#>         1         2         3
f(bce=10)
#>  FirstArg SecondArg  ThirdArg
#>         1         2        10
f(30,40)
#>  FirstArg SecondArg  ThirdArg
#>        30        40         3
f(bce=20,abc=10,100)
#>  FirstArg SecondArg  ThirdArg
#>        10       100        20
```

Hadley Wickham (Wickham, 2019) provides valuable advice for passing arguments to functions, for example: (1) to focus on positional mapping for the first one or two arguments, (2) to avoid positional mapping for arguments that are not used too often, and (3) that unnamed arguments should come before named arguments.

A final argument worth exploring is the ... argument, which will match any arguments not otherwise matched, and this can be easily forwarded to other functions, or also explored, by converting to a list in order to examine the arguments passed. Here is an example of how it can be used to pass any number of arguments to a function, and how the function can access these arguments.

```
test_dot1 <- function(...){
  ar = list(...)
  str(ar)
}
test_dot1(a=10,b=20:21)
#> List of 2
#>  $ a: num 10
#>  $ b: int [1:2] 20 21
```

However, the ... argument is often used to forward a set of arguments to another function; for example, here we can see how we can add flexibility to the function `test_dot2` by adding the argument ... as a parameter.

```
test_dot2 <- function(v,...){
  sum(v,...)
}
v <- c(1:3,NA)
test_dot2(v)
#> [1] NA
test_dot2(v,na.rm=TRUE)
#> [1] 6
```

## 4.4  Error checking for functions

While functions are invaluable as small units of useful code, they must also be robust. When an error is encountered, it should be dealt with. From a programming perspective, a decision needs to be made as to whether an error condition requires that the program be halted, or whether an error generates information that can be relayed to the user. We will take the first approach in this short example, and assume that the program must stop when an error is encountered. This can be done using R's `stop()` function, which stops execution of the current expression, and executes an error action. The general process for creating robust functions is to test conditions early in the function, and so "fail fast" (Wickham, 2019).

Here, we return to our first example for extracting even values from a vector. For this case, we need to think about how we intend the function to be called; for example, what should happen if the input vector:

- Is empty?
- Is not an atomic vector?
- Is not numeric?

These would seeem like sensible checks to make before we would proceed with the function's core processing. In order to test whether or not a vector is empty, we can use the `length()` function. For example:

```
v <- c() # an empty vector
length(v) == 0
#> [1] TRUE
```

Next, we need to make sure the vector is a numeric vector; for example, if a user sent in a character vector, it would not be possible to use the %% operator on a character vector.

```
v <- c("Hello", "World")
is.numeric(v)
#> [1] FALSE
```

Therefore, these two functions can be used to initially check the input values, and "fail fast" if necessary. We add this checking logic to the function's opening lines.

```
evens <- function(v){
  if(length(v)==0)
    stop("Error> exiting evens(), input vector is empty")
  else if(!is.numeric(v))
    stop("Error> exiting evens(), input vector not numeric")
  v[v%%2==0]
}
```

```
# Robustness test 1, check for empty vector
t1 <- c()
evens(t1)
# Error in evens(t1) : Error> exiting evens(), input vector is empty
```

```
# Robustness test 2, check for non-numeric vector
t2 <- c("This should fail")
evens(t2)
# Error in evens(t2) : Error> exiting evens(), input vector not numeric
```

```
# Robustness test 3, check for non-atomic vector
t3 <- list(1:10)
evens(t3)
# Error in evens(t3) : Error> exiting evens(), input vector not numeric
```

```
# Robustness test 4, should work ok
t4 <- 1:7
evens(t4)
#> [1] 2 4 6
```

## 4.5   Environments and functions

Understanding how environments work is key to figuring out how variables are accessed in R. It is worth spending time on understanding an environment, which comprises two parts: (1) a frame that contains name-object bindings, and (2) a reference to its parent environment. This reference mechanism creates a hierarchy of environments within R. We will come back to this idea of the environment hierarchy shortly, but before that, we will consider the global environment $R\_GlobalEnv$, which is the interactive workspace that contains user-defined variables and functions. In showing the examples, we will make use of the package `pryr`, and in particular, the function `where()`, which, for any given R object - expressed as a string - will display the environment where it is located.

Consider the following example, where we assume that initially the global environment does not contain any variables. We define three variables x, y and z.

```r
library(pryr)
x <- c(1,2,3)
y <- c(3,4,6)
z <- x * y
pryr::where("x")
#> <environment: R_GlobalEnv>
pryr::where("y")
#> <environment: R_GlobalEnv>
pryr::where("z")
#> <environment: R_GlobalEnv>
```

As we can see, all of the variables defined are contained in the global environment. Note that the R function `globalenv()` can also return a reference to the global environment. This can be seen by calling the function `ls()` which returns a vector of character strings, providing the names of the objects in the specified environment.

```r
ls(envir=globalenv())
#> [1] "x" "y" "z"
```

We can visualize an environment and its variables, and an example of this is shown in Figure 4.1. While a detailed technical discussion of how variables are stored in R is outside the scope of this text, the main idea is captured, namely:

- The values of a variable are stored in memory, and for a vector, these are in successive (i.e., contiguous) locations. We can visualize this storage as an array structure of three cells for each of the variables x, y, and z.
- We also need a variable name (identifier) that references (or "points to") the values in memory, and the link between the variable and its value in memory is known as a *binding*. If we did not define a variable to "point to" the memory location, we could not retrieve or manipulate the stored data.

Environments are also important for understanding how functions work. When a function is created, it obtains a reference (i.e., it "points to") the environment in which it was created, and this is known as the function's *enclosing environment*. The function `environment()` can be used to confirm a function's enclosing environment. Consider the following code for the function `f1()`. It

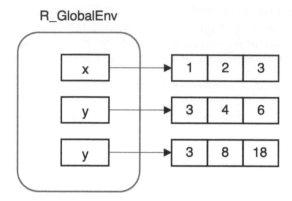

**FIGURE 4.1** Visualizing how variables are part of an environment

takes in two arguments a and b, sums these, and then multiplies the answer by z. However, how does the function find the value for z?

```r
f1 <- function(a,b){
  (a+b)*z
}
environment(f1)
#> <environment: R_GlobalEnv>
```

The answer to this important question is visualized in Figure 4.2, which highlights a key aspect of R's scoping rules, namely, the way in which R can access variables. A function is also stored in memory, and it is referenced by a variable, in this case f1 that binds to the function. We can show that the function also accepts two arguments, a and b.

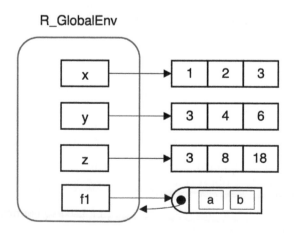

**FIGURE 4.2** Adding a function to the environment

But the diagram also shows an interesting feature of R, in that the function also contains a reference to its enclosing environment. This means that when the function *executes*, it also has a pathway to search its enclosing environment for variables and functions. For example, we can see that the function has the equation (a+b)*z, where a and b are local variables, and so are already part of the function. However, z is not part of the function, and therefore R will search the enclosing environment for z, and if it finds it, will use that value in the calculation.

This is shown below. Note that if z was not defined in the global environment, this function would generate an error.

```
z
#> [1]   3   8 18
ans <- f1(2,3)
ans
#> [1] 15 40 90
```

Therefore, functions can access variables in their enclosing environment, which in this example is the global environment. While the function f1() can read the variable z, it cannot change the value of z using the <- operator.

However, in R there is another operator known as the *superassignment operator* <<-, and this can be used within functions to modify a variable in the parent environment. If the variable does not exist in the parent environment, then a variable will be created in the global environment.

To show the idea, consider the following function f2().

```
c <- 20
f2 <- function(a,b){
  ans <- a + b + c
  c <<- 100
  ans
}
```

We now call f2() and show the value of c. Note that within the function f2(), the superassignment operator has changed the value in the global environment. We will not use the superassignment operator in the textbook; however, it is informative to know that this feature is available in R. The operator tends to be used for ideas such as function factories (Wickham, 2019), where a function can return functions and also maintain shared state variables (these are known as *closures*, but are outside the scope of this book).

```
c
#> [1] 20
d <- f2(2,4)
d
#> [1] 26
```

```
c
#> [1] 100
```

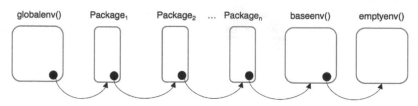

**FIGURE 4.3** Environment structure in R

We now come to an important diagram in R, which is shown in Figure 4.3. This should help clarify the mechanisms by which R accesses variables. Essentially, environments form a tree structure, in which every environment has a parent environment, apart from the environment at the top of the tree which is known as the empty environment (accessed through the function `emptyenv()`). Many of R's functions are stored in the base environment (accessed using the function `baseenv()`), for example, functions such as `min()` and `max()`. Their location can be confirmed using the `where()` function.

```
where("min")
#> <environment: base>
where("max")
#> <environment: base>
```

What is interesting is that a separate environment is also added for each new package loaded using `library()`, and the newest package's environment is added as the direct parent of the global environment. The full hierarchy can be easily shown using the function `search()`. Notice that the first environment shown is always the global environment, and the last environment shown is the base environment, as `search()` does not show the empty environment, although the empty environment can be shown through a call to the function `parent.env()`.

```
# Show the empty environment
parent.env(baseenv())
#> <environment: R_EmptyEnv>
```

Given this structure, we can now describe the process that R follows in order to evaluate an expression. Consider the following:

```
ans <- max(x)
ans
#> [1] 3
```

In this case, R starts in the global environment, and locates the variable x. It then will search for the function `max()`. As `max()` is not defined in the global environment, R will locate the reference to the parent environment, and search that environment for the function. It will then iterate through the entire hierarchy until it reaches the base environment. There it will find the function `max()`, call it, and return the result.

To further clarify this process that R uses to find the variable, we can write a new function called `max()` which will be located in the global environment, and we note what happens when we now call `max()`.

```
# define a function called min
max <- function(v){
  "Hello World"
}

ans <- max(x)
ans
#> [1] "Hello World"
```

Clearly, this is not the result we might have expected, and we could protest that R has called the new `max()` instead of the obvious function in the base environment. However, this is just the way the R evaluation process works (we have to accept this!), as it starts in the global environment and finds and executes the first matching function. Interestingly, R provides a safe mechanism to call the exact function you want by prefixing the call to indicate the environment name, for example, `base::max()`, which calls the `max()` function defined in the base environment. We can see the overall process below. Note that at any time we can remove a variable from an environment, by using the function `rm()` and passing in the variable name as a character vector.

```
max(x)
#> [1] "Hello World"
base::max(x)
#> [1] 3
rm("max")
```

## 4.6 Functionals with `lapply()`

In the previous chapter, we demonstrated how the for loop can be used to iterate over a list, element by element. We now introduce an important aspect of programming with R, which is the use of functionals, which take functions as part of their input, and use that function to process data. In many cases, functionals can be used instead of loops to iterate over data and return a result. One of the most important functionals that can be used is `lapply(x, f)`, which:

- Accepts as input a list x and a function f,
- Returns as output a new list of exactly the same length as x, where each element in the new list is the result of applying the function f to the corresponding element of the input list x.

We show a simple implementation of this idea (Wickham, 2019), and will name this function my_lapply(x,f).

```r
my_lapply <- function(x,f){
  # Create the output list vector
  o <- vector(mode="list",length = length(x))
  # Loop through the entire input list
  for(i in seq_along(x)){
    # Apply the function to each element and
    # store in the corresponding output location
    o[[i]] <- f(x[[i]])
  }
  # Return the output list
  o
}


l_in <- list(1:4,11:14,21:24)
l_out <- my_lapply(l_in,mean)
str(l_out)
#> List of 3
#>  $ : num 2.5
#>  $ : num 12.5
#>  $ : num 22.5
```

The process followed is:

- Two inputs are passed to the function, the data (a list of three elements), and the function to apply to the elements, which in this example is the function mean().
- Inside the function, the list l_in is mapped to the variable x, and the function mean is mapped to f.
- The first action is to create a variable (o) that will eventually store the result. We know that the number of elements in this output variable must be the same as the number of elements on the input list. That's the key idea behind the lapply() functional; it operates on each element with the same function, and returns all the results.
- The function then iterates over the entire input list (now stored in x), and calls the function f with this data, storing the result of this calculation in the corresponding location of the variable o.
- Once all the list elements have been processed, the variable o is returned, which is a list of three elements, and each element contains the mean of the corresponding input list element.

So while it's instructive to see how this function works, there is no need to duplicate the code by creating your own version. The code below shows the solution using `lapply()`.

```
l_in <- list(1:4,11:14,21:24)
l_out <- lapply(l_in,mean)
str(l_out)
#> List of 3
#>  $ : num 2.5
#>  $ : num 12.5
#>  $ : num 22.5
```

In later chapters, when we introduce the `purrr` package, which provides a set of functions that can be used to iterate over data structures, in a similar way to `lapply()`, but with more functionality.

## 4.7 Mini-case: Star Wars movies (revisited) using functionals

Here we revisit the examples from Chapter 3, and the sole purpose of these code snippets is to demonstrate how the loop structures previously used can now be replaced by the function `lapply()`. The library `repurrrsive`, which contains great examples of lists, is used, in particular, the examples based on the Star Wars movies.

Our first example of replacing loops with `lapply()` is shown below. To remind ourselves, the goal is to find the movies directed by George Lucas. We will use `lapply()` to identify which elements are relevant, and in this solution, `lapply()` takes two arguments:

- The list `sw_films`, containing seven elements, where each element is itself a list of 14 elements, one of which is the movie director.
- An anonymous function that takes in each successive list element (parameter x), and compares the `$director` element with the variable `target`, which is defined in the global environment, and so can be accessed by the anonymous function. The anonymous function returns the last evaluated expression, which is the result of the relational expression (either `TRUE` or `FALSE`).

The `lapply()` call returns a list of seven elements, each of which contains a logical vector. This list can be converted to an atomic vector using the function `unlist()`, and the output is then used to filter the original list, which is then stored in the variable `target_list`.

```
library(repurrrsive)
# Search for movies by George Lucas and store these in a new list
```

```
target        <- "George Lucas"

# Call lapply to return a list of logical vectors
is_target    <- lapply(sw_films,function(x)x$director==target)

# Convert this list to an atomic vector, which is needed for filtering
is_target    <- unlist(is_target)

# Filter the list to contain the George Lucas movies
target_list <- sw_films[is_target]
length(target_list)
#> [1] 4
```

With the flexibility provided by anonymous functions, we can design an alternative way to filter the list. This is a common theme in R, often there are many ways to arrive at the same outcome. The code inside the anonymous function now uses an if statement to return the actual list element (x) if the condition is true, and NA otherwise. Therefore, this lapply() call will return a list of seven, although all those films that do not have George Lucas as a director will be replaced by NA. This is convenient, because the function is.na() can then be used to filter the list, as is shown below in the code example.

```
# Search for movies by George Lucas and store these in a new list
target        <- "George Lucas"
target_list <- lapply(sw_films,function(x)
                                 if(x$director==target) x
                                 else NA)
target_list <- target_list[!is.na(target_list)]
length(target_list)
#> [1] 4
```

With the filtered list, you can then call lapply() to return the movie titles.

```
# Get the movie titles as a list
movies <- lapply(target_list,function(x)x$title)
movies <- unlist(movies)
movies
#> [1] "A New Hope"           "Attack of the Clones"
#> [3] "The Phantom Menace"   "Revenge of the Sith"
```

Using lapply(), we can also create a new list from the original list. Here we create a list with three elements, where each element will store all of the data for the title, episode number and movie director.

```
# Create a new list to store the data in a different way
sw_films1              <- list(title=c(), episode_id=c(), director=c())
```

```
sw_films1$title       <- unlist(lapply(sw_films,
                                       function(x)x$title))

sw_films1$episode_id  <- unlist(lapply(sw_films,
                                       function(x)x$episode_id))

sw_films1$director    <- unlist(lapply(sw_films,
                                       function(x)x$director))
```

Given this new structure, we can then extract the movies for George Lucas using the following subsetting operation. This works because the $ operator extracts the contents of the list element `title`, and then we simply subset this as an atomic vector.

```
# Find all the movie titles by George Lucas
sw_films1$title[sw_films1$director=="George Lucas"]
#> [1] "A New Hope"           "Attack of the Clones"
#> [3] "The Phantom Menace"   "Revenge of the Sith"
```

We can use `lapply()` to find all the planet diameters in the list `sw_planets`, replacing the string "unknown" with NA, and then summarize the vector, showing the overall statistics for this variable.

```
# Return all planet diameters as numeric, with NA where it's unknown
diameters <- unlist(
            lapply(sw_planets,
                   function(x)
                     if (x$diameter != "unknown")
                       as.numeric(x$diameter)
                     else NA))
# Provide a summary of the data
summary(diameters)
#>    Min. 1st Qu.  Median    Mean 3rd Qu.    Max.    NA's
#>       0    7812   11015   12388   13422  118000      17
```

Finally, we can extract the passenger capacity for all starships contained in the list `sw_starships`.

```
# Get the passenger capacity for all spaceships
passengers <- unlist(lapply(sw_starships,
                     function (x)
                       if (x$passengers != "unknown")
                         as.numeric(x$passengers)
                       else NA))
# Show the data
passengers
#> [1]      75 843342       6       0       0       0   38000       6      20
```

```
#> [10]      75    1200       0       0      16       0      NA       6      10
#> [19]   30000       0      NA       0   48247      16      NA      90  139000
#> [28]   16000      11    2000       3       0       0      NA       0       0
#> [37]     600
```

In summary, an important point is that we need to be ready to process any kind of list structure, and the function `lapply()` is helpful to organize whatever task we may face.

## 4.8    Creating a data processing pipeline using R's native pipe operator

The native pipe operator in R, represented by the symbol `|>`, allows you to chain a number of operations together, without having to assign intermediate variables. Ths operator, originally based on the `%>%` operator from the package `magrittr` (Bache and Wickham, 2014), allows you to construct a data processing pipeline, where an input is identified (e.g., a list, vector, or later in the book, a data frame), an output specified, and each discrete step in generating the output is linked together in a chain. The key idea is that the output from one step becomes the input to the second step, and it provides an elegant way to assemble the different steps.

The general format of the pipe operator is LHS `|>` RHS, where LHS is the first argument of the function defined on the RHS. For example, let's say we want to get the minimum of a randomized vector value (we will use the function `runif()` for this, which generates a random uniformly distributed number in the range $0 - 1$.)

With the `|>` operator, we first identify the LHS data for `min()`, and then we insert the function `min()` with no arguments as the RHS of the expression.

```
set.seed(200)
# Generate a vector  of random numbers
n1 <- runif(n = 10)
# Show the minimum the usual way
min(n1)
#> [1] 0.0965
# Use the native pipe to generate the same answer
n1 |> min()
#> [1] 0.0965
```

More operations can be added to the chain, and in that case, the output from the first RHS then becomes the LHS for the next operation. For example, we could also add an addition operation to the example, to round the number of decimal places to 3, by using the round() function.

```
n1 |> min() |> round(3)
#> [1] 0.097
```

As mentioned, lists are frequently used as a starting point, and we will use the data frame mtcars (we will present data frames in Chapter 5) to perform the following chain of data transformations:

- Take as input mtcars.
- Convert mtcars to a list, using the function as.list(). Note that data frames are technically a list.
- Process the list one element at a time, and get the average value of each variable.
- Convert the list returned by lapply() to an atomic vector (using unlist()).
- Store the result in a variable.

The full instruction is shown below.

```
a1 <- mtcars                       |>   # The input
      as.list()                    |>   # Convert to a list
      lapply(function(x)mean(x))   |>   # Get the mean of each element
      unlist()                          # Convert to atomic vector
a1
#>       mpg      cyl      disp       hp      drat       wt      qsec
#>   20.0906   6.1875  230.7219  146.6875   3.5966   3.2172   17.8487
#>        vs       am      gear     carb
#>    0.4375   0.4062    3.6875    2.8125
```

A number of points are worth noting:

- While not essential from a computation point of view, it is good practice to start a new line for each step.
- Adding short descriptive comments, with the comment sign #, makes the code easier to understand.

## 4.9 Summary of R functions from Chapter 4

Here is a summary of the new functions we introduced in this chapter. Note that calls to functions contained in libraries that are not part of base R require that you first load the library using the function library().

| Function | Description |
|----------|-------------|
| duplicated() | Identifies duplicates in a vector. |
| where() | Returns the environment for a variable (pryr library). |
| environment() | Finds the environment for a function. |
| library() | Loads and attaches add-on packages. |
| parent.env() | Finds the parent environment for an input environment. |
| search() | Returns a vector of environment names. |
| globalenv() | Returns a reference to the global environment. |
| baseenv() | Returns a reference to the base environment. |
| lapply(x,f) | Applies f to each element of x and returns results in a list. |
| rm() | Removes an object from an environment. |
| stop() | Stops execution of the current expression. |
| unique() | Returns a vector with duplicated elements removed. |

## 4.10    Exercises

1. Write a function `get_even1()` that returns only the even numbers from a vector. Make use of R's modulus function `%%` as part of the calculation. Try to implement the solution as one line of code. The function should transform the input vector in the following way.

```
set.seed(200)
v <- sample(1:20,10)
v
#>  [1]  6 18 15  8  7 12 19  5 10  2
v1 <- get_even1(v)
v1
#>  [1]  6 18  8 12 10  2
```

2. Write a similar function `get_even2()` that takes a second parameter `na.omit`, with a default of `FALSE`. If `na.omit` is set to `TRUE`, the vector is pre-processed in the function to remove all `NA` values before doing the final calculation.

```
set.seed(200)
v <- sample(1:20,10)
i <- c(1,5,7)
v[i] <- NA
v
#>  [1] NA 18 15  8 NA 12 NA  5 10  2
v1 <- get_even2(v)
```

```
v1
#> [1] NA 18   8 NA 12 NA 10   2
v2 <- get_even2(v,na.omit=TRUE)
v2
#> [1] 18  8 12 10  2
```

3. Which one of the following three functions calls to `fn_test()` will not work. Why?

```
# The function
fn_test <- function(a, b, c){
  a+b
}
# Call 1
fn_test(1,2)
# Call 2
fn_test(c=1,2)
# Call 3
fn_test(b=1,10)
```

4. What will be the output from the following function call?

```
a <- 100

env_test <- function(b,c=20){
  a+b+c
}

env_test(1)
```

5. Use `lapply()` followed by an appropiate post-processing function call, to generate the following output, based on the input list.

```
# Create the list that will be processed by lapply
l1 <- list(a=1:5,b=100:200,c=1000:5000)

# The result is stored in ans
ans
#>    a    b    c
#>    3  150 3000
```

# 5

## Matrices and Data Frames

On an intuitive level, a *data frame* is like a matrix, with a two-dimensional rows-and-columns structure. However, it differs from a matrix in that each column may have a different type.

— Norman Matloff (Matloff, 2011)

## 5.1 Introduction

To date, we have used atomic vectors and lists to store information. While these are foundational data structures in R, they do not provide support for processing *rectangular data*, which is a common format in data science. Rectangular data, as the name suggests, is the usual format used in spreadsheets or databases, and it comprises rows of data for one or more variables. Typically, in a rectangular dataset, every column represents a variable (where each variable can have a different type), and each row contains an observation, for example, a set of values that are related (medical data on a single patient, or weather data at a specific point in time). More generally, we can have two types of rectangular (two-dimensional) data: (1) data of a same type, typically numeric, that is stored in a matrix, and (2) data of different types that is stored in a data frame. This chapter introduces the matrix and the data frame, and shows how we can process data in these structures, using ideas from subsetting atomic vectors and lists.

Upon completing the chapter, you should understand:

- How to create a matrix in R, using a one-dimensional vector as input.
- How to subset a matrix, and extend it by adding rows and columns.

- How to create a data frame, and how to subset the data frame using matrix notation.
- How to create a tibble, and understand how a tibble differs from a data frame.
- How to use the R functions `subset()` and `transform()` to process data frames.
- How to use the function `apply()` to process matrices and data frame, on either a row or a column basis.
- How to manipulate matrices and data frames using base R functions.
- How to solve all five test exercises.

### Chapter structure

- Section 5.2 introduces the matrix, and shows how it can be created using the function `matrix()`.
- Section 5.3 presents the data frame, which is also a list.
- Section 5.5 summarizes the tibble, which is an updated version of the data frame with a number of additional features.
- Section 5.6 shows how functionals can be applied to data frames and matrices, in particular the function `apply()`, which can process either rows or columns in a matrix and a data frame.
- Sections 5.7 and 5.8 present the chapter's two *mini-cases*. The first shows how a matrix can be used to store information on a synthetic social network structure, and that can capture connections between individuals; while the second mini-case highlights how a pipeline can be set up to process a data frame.
- Section 5.9 provides a summary of all the functions introduced in the chapter.
- Section 5.10 provides a number of short coding exercises to test your knowledge.

## 5.2   Matrices

In R, a matrix is a two-dimensional structure, with rows and columns, that contains the same type. A simple way to understand a matrix is that it is an atomic vector in two dimensions, and is created using the `matrix()` function, with the following arguments:

- `data`, which are the initial values, contained in an atomic vector, supplied to the matrix,
- `nrow`, the desired number of rows,
- `ncol`, the desired number of columns,
- `byrow`, a logical value (default is FALSE), that specifies what way to fill the matrix with data, either filled by row or by column,
- `dimnames`, a list of length 2 giving row and column names, respectively.

We now use this information to create a 3-by-3 matrix.

```
set.seed(100)
data <- sample(1:9)
data
#> [1] 7 6 3 1 2 5 9 4 8
m1 <- matrix(data,
             nrow = 3,
             ncol=3,
             dimnames = list(c("R1","R2","R3"),
                             c("C1","C2","C3")))
# Display the matrix
m1
#>    C1 C2 C3
#> R1  7  1  9
#> R2  6  2  4
#> R3  3  5  8

# Show the number of rows and columns
nrow(m1)
#> [1] 3
ncol(m1)
#> [1] 3

# Show the matrix dimensions
dim(m1)
#> [1] 3 3
```

A number of points that are worth noting are:

- The matrix is populated by column order as the default. If by_row was set to TRUE, then the matrix would be populated by row order.
- The row and column names are set using the dimnames argument. This is not required, and row names and column names can always be set on a matrix using the functions rownames() and colnames().
- The functions nrow() and ncol() can be used to return the matrix dimensions.
- The function dim() provides information on the matrix dimensions, and can also be used to resize a matrix, for example, converting a $3 \times 3$ to a $1 \times 9$.

An important property of a matrix is that it can be extended, by adding either rows, using the function rbind(); or columns, with the function cbind(). Here is an example of how we can extend the original $3 \times 3$ array by adding a row, and then naming that row "R4".

```
m1_r <- rbind(m1,c(1,2,3))
rownames(m1_r)[4] <- "R4"
m1_r
```

```
#>     C1 C2 C3
#> R1   7  1  9
#> R2   6  2  4
#> R3   3  5  8
#> R4   1  2  3
```

In a similar way, we can add a new column to the matrix using cbind(), and also name this new column using the function colnames().

```
m1_c <- cbind(m1_r,c(10,20,30,40))
colnames(m1_c)[4] <- "C4"
m1_c
#>     C1 C2 C3 C4
#> R1   7  1  9 10
#> R2   6  2  4 20
#> R3   3  5  8 30
#> R4   1  2  3 40
```

To subset a matrix, we supply an index for each dimension [row number, column number]; for example, here we extract the value contained in row 2, column 2.

```
m1
#>     C1 C2 C3
#> R1   7  1  9
#> R2   6  2  4
#> R3   3  5  8

# Extract the value in row 2, column 2
m1 [2,2]
#> [1] 2
```

Ranges can also be supplied during subsetting; for example, the first two rows and the first two columns.

```
m1[1:2,1:2]
#>     C1 C2
#> R1   7  1
#> R2   6  2
```

Blank subsetting for one of the dimensions lets you retain all of the rows, or all of the columns. Note that if only one column is subsetted, *R will return a vector by default.* To keep the matrix structure, the argument drop=FALSE is added.

```
# Extract first row and all columns
m1[1,]
#> C1 C2 C3
```

```
#>  7  1  9
```

```
# Extract first column, returned as a vector
m1[,1]
#> R1 R2 R3
#>  7  6  3
```

```
# Extract all rows and the first column, returned as matrix
m1[,1,drop=FALSE]
#>    C1
#> R1  7
#> R2  6
#> R3  3
```

Overall, the subsetting rules we applied to an atomic vector also apply to a matrix, for example, subsetting by row and column name.

```
# Extract first row and first two columns
m1["R1",c("C1","C2")]
#> C1 C2
#>  7  1
# Extract all rows and first two columns
m1[,c("C1","C2")]
#>    C1 C2
#> R1  7  1
#> R2  6  2
#> R3  3  5
```

Logical vectors can also be used to subset matrices; for example, to extract every second row from the matrix, the following code will work.

```
m1
#>    C1 C2 C3
#> R1  7  1  9
#> R2  6  2  4
#> R3  3  5  8
m1[c(T,F),]
#>    C1 C2 C3
#> R1  7  1  9
#> R3  3  5  8
```

There are important functions available to process matrices, and the main ones are now summarized.

- The function is.matrix() is used to test whether the object is a matrix or not. This can be used as a pre-test, in order to ensure that the data is in the expected format.

```
A <- matrix(1:4,nrow=2)
B <- matrix(1:4,nrow=2,byrow = T)
C <- list(c1=1:2, c2=3:4)
is.matrix(A)
#> [1] TRUE
is.matrix(B)
#> [1] TRUE
is.matrix(C)
#> [1] FALSE
```

- By default, matrix operations such as multiplication, division, exponentiation, addition, and subtraction are performed on an element-wise basis. For example, we now explore some simple matrix operations, including addition and multiplication.

```
A <- matrix(1:4,
            nrow=2,
            dimnames = list(c("A_R1","A_R2"),c("A_C1","A_C2")))
B <- matrix(1:4,
            nrow=2,
            byrow = T,
            dimnames = list(c("B_R1","B_R2"),c("B_C1","B_C2")))
A
#>      A_C1 A_C2
#> A_R1    1    3
#> A_R2    2    4


B
#>      B_C1 B_C2
#> B_R1    1    2
#> B_R2    3    4


# Multiplication of A and B
A*B
#>      A_C1 A_C2
#> A_R1    1    6
#> A_R2    6   16


# Addition of A and B
A+B
#>      A_C1 A_C2
#> A_R1    2    5
#> A_R2    5    8


# Multiplying A by a constant
```

```
10*A
#>       A_C1 A_C2
#> A_R1    10    30
#> A_R2    20    40
```

- There is a special operator %*% that is used for matrix multiplication.

```
# Use matrix algebra to multiply two matrices
A%*%B
#>       B_C1 B_C2
#> A_R1    10    14
#> A_R2    14    20
```

- Use the function t() to obtain the transpose of a matrix.

```
t(A)
#>       A_R1 A_R2
#> A_C1     1     2
#> A_C2     3     4
```

- The function dim() can be used to return the matrix dimensions, namely, the number of rows and the number of columns.

```
dim(A)
#> [1] 2 2
```

- The function dimnames() returns the row and column names in a list.

```
dimnames(A)
#> [[1]]
#> [1] "A_R1" "A_R2"
#>
#> [[2]]
#> [1] "A_C1" "A_C2"
```

- The functions rownames() and colnames() return the row names and column names as atomic vectors.

```
rownames(A)
#> [1] "A_R1" "A_R2"
```

```
colnames(A)
#> [1] "A_C1" "A_C2"
```

- The function diag() can be used in two ways. First, it can set all the diagonal elements of a matrix. Second, it can be used to create the identity matrix for a given dimension.

```
diag(B) <- -1
B
```

```
#>        B_C1 B_C2
#> B_R1   -1    2
#> B_R2    3   -1

I <- diag(3)
I
#>       [,1] [,2] [,3]
#> [1,]    1    0    0
#> [2,]    0    1    0
#> [3,]    0    0    1
```

- The function eigen() can be used for calculating the eigenvalues and eigenvectors of a matrix, which has important applications in many engineering and computing problems. Here is an example of a $2 \times 2$ matrix, where both eigenvalues and eigenvectors are shown.

```
M <- matrix(c(1,-1,-2,3),nrow=2)
M
#>       [,1] [,2]
#> [1,]    1   -2
#> [2,]   -1    3

eig <- eigen(M)

eig$values
#> [1] 3.7321 0.2679
eig$vectors
#>             [,1]     [,2]
#> [1,]  0.5907 -0.9391
#> [2,] -0.8069 -0.3437
```

- The determinant of a matrix is calculated using the function det().

```
det(M)
#> [1] 1
```

- The functions colSums(), colMeans(), rowSums() and rowMeans() can be used to generate summary statistics for a matrix.

```
rowSums(A)
#> A_R1 A_R2
#>    4    6
rowMeans(A)
#> A_R1 A_R2
#>    2    3
colSums(A)
#> A_C1 A_C2
```

```
#>      3     7
colMeans(A)
#> A_C1 A_C2
#>  1.5  3.5
```

In summary, R provides good support for problems that require matrix manipulation and matrices need to be defined using the `matrix()` function. Many of the subsetting commands used for atomic vectors can also be used for matrices, and that includes referencing elements by the row/column name. Matrices can be extended easily, using functions such as `cbind()` and `rbind()`.

Furthermore, functions such as `diag()`, `eigen()` and `det()` can be used to support analysis. However, an important feature of matrices is that all the values need to be the same type, and while this is suitable for the examples shown, there are cases where rectangular data must store data of different types, for example, the type of data contained in a typical dataset. For these situations, the *data frame* is an ideal mechanism to store, and process, heterogeneous data.

## 5.3   Data frames

On an intuitive level, a data frame is similar to a matrix, with a two-dimensional row and column structure, while on a technical level, a data frame is a list, with the elements of that list containing equal length vectors (Matloff, 2011).

Crucially, the elements (columns) of a data frame can be of different types, and this feature provides functionality for processing datasets as part of the operations research process. The data frame, with its row and column structure, will be familiar to anyone who has used a spreadsheet, where each column is a variable, and every row is an observation (Lander, 2017). The data frame, and its successor, the *tibble*, will be used extensively in Parts II and III of this book.

In most cases, a data frame is created by reading in a rectangular data structure from a file or a database. For example, the function `read.csv()` can be used to read in a file and create a data frame. However, a data frame can also be created using the function `data.frame()`, along with a set of equally sized vectors. For example, we could set up a simple data frame with five rows and three columns. Note that we set the argument `stringsAsFactors` to be `FALSE`, so that the second column contains characters, rather than factors.

```
d <- data.frame(Number=1:5,
                Letter=LETTERS[1:5],
                Flag=c(T,F,T,F,T),
                stringsAsFactors = F)
```

```
d
#>    Number Letter  Flag
#> 1       1      A  TRUE
#> 2       2      B FALSE
#> 3       3      C  TRUE
#> 4       4      D FALSE
#> 5       5      E  TRUE
```

As an aside, a factor is an R object (it's actually an integer type with additional S3 class attributes, discussed in Chapter 6) that is used to store categories, and can be created with a call to the R function factor(). For example, we could use the sample() function to generate random categories for different age groups, convert those to a factor, and then summarize the values. Note the additional information that is displayed when we use a factor instead of a character.

```
set.seed(100)
vals    <- sample(c("Age0-15",
                    "Age16-64",
                    "Age65+"), 10, replace = T)
f_vals <- factor(vals)
summary(f_vals)
#>  Age0-15 Age16-64   Age65+
#>        1        6        3
```

When exploring data frames, the summary() function provides a convenient synopsis for each variable (or column) in the data frame.

```
summary(d)
#>       Number      Letter              Flag
#>  Min.   :1   Length:5          Mode :logical
#>  1st Qu.:2   Class :character  FALSE:2
#>  Median :3   Mode  :character  TRUE :3
#>  Mean   :3
#>  3rd Qu.:4
#>  Max.   :5
```

An important activity that is required with a data frame is to be able to: (1) subset rows, (2) subset columns, and (3) add new columns. Because a data frame is a list and also shares properties of a matrix, we can combine subsetting mechanisms from both of these data structures to subset a data frame. Given that a data frame is also a list, we can access a data frame column using the $ operator. A number of subsetting examples are now presented.

- To select the first two rows from the data frame.

```
d[1:2,]
#>    Number Letter  Flag
```

```
#> 1         1         A    TRUE
#> 2         2         B FALSE
```

- To select all rows that have the column Flag set to true.

```
d[d$Flag == T,]
#>    Number Letter Flag
#> 1       1      A TRUE
#> 3       3      C TRUE
#> 5       5      E TRUE
```

- To select the first two rows and the last two columns.

```
d[1:2,c("Letter","Flag")]
#>    Letter  Flag
#> 1       A  TRUE
#> 2       B FALSE
```

- To add a new column, we can simply add a new element as we would with a list.

```
d1 <- d
d1$letter <- letters[1:5]
d1
#>    Number Letter  Flag letter
#> 1       1      A  TRUE      a
#> 2       2      B FALSE      b
#> 3       3      C  TRUE      c
#> 4       4      D FALSE      d
#> 5       5      E  TRUE      e
```

- To add a new column, we can also use the cbind() function.

```
d2 <- cbind(d,letter2=letters[6:10])
d2
#>    Number Letter  Flag letter2
#> 1       1      A  TRUE       f
#> 2       2      B FALSE       g
#> 3       3      C  TRUE       h
#> 4       4      D FALSE       i
#> 5       5      E  TRUE       j
```

## 5.4   R functions for processing data frames: `subset()` and `transform()`

While subsetting and adding new elements to data frames can be performed using the methods just summarized, there are special purpose functions that make the process easier and more intuitive.

The function `subset(x, subset,select)` returns subsets of vectors, matrices, or data frames that meet specified conditions. The main arguments to provide when subsetting data frames are:

- `x`, the object to be subsetted,
- `subset`, a logical expression indicating which rows should be kept,
- `select`, which indicates the columns to be selected from the data frame. If this is not present, all columns are returned.

Here is an example of subsetting the data frame `mtcars`, where all cars with an `mpg` greater than 32 are filtered, and only two of the columns returned.

```
subset(mtcars,mpg>32,select=c("mpg","disp"))
#>                mpg disp
#> Fiat 128      32.4 78.7
#> Toyota Corolla 33.9 71.1
```

Of course, we could achieve the same outcome without using `subset()`, and utilize the matrix-type subsetting for data frames, or by using the `$` operator. However, even though the same result is achieved, the `subset()` function makes it easier for others to understand the code.

```
mtcars[mtcars[,"mpg"]>32,c("mpg","disp")]
#>                mpg disp
#> Fiat 128      32.4 78.7
#> Toyota Corolla 33.9 71.1
mtcars[mtcars$mpg>32,c("mpg","disp")]
#>                mpg disp
#> Fiat 128      32.4 78.7
#> Toyota Corolla 33.9 71.1
```

A second function that can be used to manipulate data frames is `transform(data, ...)`, which takes in the following arguments:

- `data`, which is the data frame,
- `...`, which are additional arguments that capture the details of how the new column is created.

We can return to our previous example, and add a new column called `kpg`, kilometers per gallon, which multiplies `mpg` by the conversion factor of 1.6. The transform() function applies the expression mpg*1.6 to each row in the data frame, and the new column can be seen in the output.

```
df1 <- subset(mtcars,mpg>32,select=c("mpg","disp"))
df1 <- transform(df1,kpg=mpg*1.6)
df1
#>                mpg disp   kpg
#> Fiat 128      32.4 78.7 51.84
#> Toyota Corolla 33.9 71.1 54.24
```

An alternative to using transform() is to use the $ operator as follows.

```
df1 <- subset(mtcars,mpg>32,select=c("mpg","disp"))
df1$kpg <- df1$mpg*1.6
df1
#>                mpg disp   kpg
#> Fiat 128      32.4 78.7 51.84
#> Toyota Corolla 33.9 71.1 54.24
```

Overall, while the functions subset() and transform() are always available to use, we will migrate to using new ways to subset and manipulate data frames and tibbles in Part *II* of the book, when we introduce the dplyr package.

## 5.5  Tibbles

As we move through the book, we will rely more on tibbles than data frames. Tibbles are a type of data frame; however they alter some data frame behaviors to make working with packages in the tidyverse easier. There are two main differences when compared to the data.frame (Wickham and Grolemund, 2016):

- Printing, where, by default, tibbles only show the first ten rows, and limit the visible columns to those that fit on the screen. The type is also displayed for each column, which is a useful feature that provides more information on each variable.

- Subsetting, where a tibble is always returned, and also partial matching are not supported.

In order to use a tibble, the tibble package should be loaded.

```
library(tibble)
#>
#> Attaching package: 'tibble'
#> The following object is masked from 'package:igraph':
#>
#>     as_data_frame
```

Similar to a data frame, a function can be used to create a tibble, and this takes a set of atomic vectors. Notice that by default string values are not converted to factors by this function, and this shows another difference with data.frame.

```
d1 <- tibble(Number=1:5,
             Letter=LETTERS[1:5],
             Flag=c(T,F,T,F,T))
d1
#> # A tibble: 5 x 3
#>    Number Letter Flag
#>     <int> <chr>  <lgl>
#> 1       1 A      TRUE
#> 2       2 B      FALSE
#> 3       3 C      TRUE
#> 4       4 D      FALSE
#> 5       5 E      TRUE
```

We can now explore some of the differences between the data.frame and tibble by comparing the two variables d and d1.

- First, we can observe their structure, using str().

```
# Show the data frame
str(d)
#> 'data.frame':    5 obs. of  3 variables:
#>  $ Number: int  1 2 3 4 5
#>  $ Letter: chr  "A" "B" "C" "D" ...
#>  $ Flag  : logi  TRUE FALSE TRUE FALSE TRUE
# Show the tibble
str(d1)
#> tibble [5 x 3] (S3: tbl_df/tbl/data.frame)
#>  $ Number: int [1:5] 1 2 3 4 5
#>  $ Letter: chr [1:5] "A" "B" "C" "D" ...
#>  $ Flag  : logi [1:5] TRUE FALSE TRUE FALSE TRUE
```

- Second, we can the see the difference in subsetting one column from each structure. Notice how the data frame output is changed to an atomic vector. In contrast, the tibble structure is retained when one column is subsetted.

```
# Subset the data frame
d[1:2,"Letter"]
#> [1] "A" "B"
# Subset the tibble
d1[1:2,"Letter"]
#> # A tibble: 2 x 1
#>    Letter
#>    <chr>
```

```
#> 1 A
#> 2 B
```

If required, it is straighforward to convert between both structures, using the function `tibble::as_tibble()` to convert from a `data.frame` to a `tibble`, and the function `as.data.frame()` to convert from a `tibble` to a `data.frame`.

```
str(as_tibble(d))
#> tibble [5 x 3] (S3: tbl_df/tbl/data.frame)
#>  $ Number: int [1:5] 1 2 3 4 5
#>  $ Letter: chr [1:5] "A" "B" "C" "D" ...
#>  $ Flag  : logi [1:5] TRUE FALSE TRUE FALSE TRUE
str(as.data.frame(d1))
#> 'data.frame':    5 obs. of  3 variables:
#>  $ Number: int  1 2 3 4 5
#>  $ Letter: chr  "A" "B" "C" "D" ...
#>  $ Flag  : logi  TRUE FALSE TRUE FALSE TRUE
```

Overall, when we make use of the `tidyverse`, which is the focus of Part II, we will mostly use tibbles as the main data structure, as their properties of maintaining the tibble structure nicely supports the chaining of data processing operations, using R's pipe operator.

---

## 5.6 Functionals on matrices and data frames

### 5.6.1 Using `apply()` on a matrix and data frame

The `apply(x,margin,f)` function is a functional used to iterate over matrices and data frames, and it accepts the following arguments:

- `x`, which can be a matrix or a data frame.
- `margin`, a number that indicates whether the iteration is by row (margin=1), or by column (margin=2).
- `f`, which is the function to be applied during each iteration.

Two examples will illustrate how the function can be used. First, we explore how `apply()` can be used on a matrix. The matrix example contains five student grades in three different subjects, where the grades are randomly generated values between 30 and 90. We also add row and column names to provide more information for the matrix variable `grades`.

```
set.seed(100)
grades <- sample(30:90,15,replace = T)
results <- matrix(grades,nrow=5)
rownames(results) <- paste0("St-",1:5)
```

```
colnames(results) <- paste0("Sub-",1:3)
results
#>        Sub-1 Sub-2 Sub-3
#> St-1     39    54    51
#> St-2     84    87    35
#> St-3     67    43    33
#> St-4     77    73    84
#> St-5     80    52    35
```

We will now use the `apply()` function to perform two tasks. First, we find the maximum grade for each subject. This involves iterating over the matrix on a column-by-column basis, and processing the data to find the maximum. Note that `apply()` simplifies the output so that, unlike `lapply()`, a numeric vector is returned instead of a list.

```
max_gr_subject <- apply(results,          # the matrix
                        2,                 # 2 for columns
                        function(x)max(x)) # the function to apply
max_gr_subject
#> Sub-1 Sub-2 Sub-3
#>    84    87    84
```

Next, we use `apply()` to find the maximum grade for each student. In this case, we need to iterate over the matrix one row at a time, and therefore the code looks like this.

```
max_gr_student <- apply(results,          # the matrix
                        1,                 # 1 for rows
                        function(x)max(x)) # the function to apply
max_gr_student
#> St-1 St-2 St-3 St-4 St-5
#>   54   87   67   84   80
```

The `apply()` function can also be used on data frames, again, to iterate and apply a function over either each row or column. For example, if we take a subset of the data frame `mtcars` and insert some random `NA` values, we can then count the missing values by either row or column.

```
set.seed(100)
my_mtcars <- mtcars[sample(1:6),c("mpg","cyl","disp")]
rows <- sample(1:nrow(my_mtcars),5)
rows
#> [1] 6 4 3 2 5

my_mtcars[rows[1],1] <- NA
my_mtcars[rows[2],2] <- NA
my_mtcars[rows[3],3] <- NA
```

```
my_mtcars[rows[4],1] <- NA
my_mtcars[rows[5],2] <- NA
my_mtcars
#>                        mpg cyl disp
#> Mazda RX1 Wag         21.0   6  160
#> Datsun 710             NA    4  108
#> Mazda RX4             21.0   6   NA
#> Valiant               18.1  NA  225
#> Hornet Sportabout     18.7  NA  360
#> Hornet 4 Drive         NA    6  258
```

First, to count the number of missing values by row, the following code can be used.

```
n_rm <- apply(my_mtcars,1,function(x)sum(is.na(x)))
n_rm
#>       Mazda RX4 Wag          Datsun 710          Mazda RX4
#>                   0                   1                  1
#>             Valiant   Hornet Sportabout     Hornet 4 Drive
#>                   1                   1                  1
sum(n_rm)
#> [1] 5
```

Second, to count the number of missing values by column, the following code can be used.

```
n_cm <- apply(my_mtcars,2,function(x)sum(is.na(x)))
n_cm
#>  mpg  cyl disp
#>    2    2    1
sum(n_cm)
#> [1] 5
```

## 5.6.2 Using `lapply()` on data frames

Given that a data frame is also a list, and that lapply() processes lists, it also means that the lapply() functional can be used to process a data frame. To explore this in more detail, we can revisit the data frame mtcars, and show its structure.

```
str(mtcars)
#> 'data.frame':    32 obs. of  11 variables:
#> $ mpg : num  21 21 22.8 21.4 18.7 18.1 14.3 24.4 22.8 19.2 ...
#> $ cyl : num  6 6 4 6 8 6 8 4 4 6 ...
#> $ disp: num  160 160 108 258 360 ...
#> $ hp  : num  110 110 93 110 175 105 245 62 95 123 ...
#> $ drat: num  3.9 3.9 3.85 3.08 3.15 2.76 3.21 3.69 3.92 3.92 ...
```

```
#>   $ wt   : num   2.62 2.88 2.32 3.21 3.44 ...
#>   $ qsec: num   16.5 17 18.6 19.4 17 ...
#>   $ vs   : num   0 0 1 1 0 1 0 1 1 1 ...
#>   $ am   : num   1 1 1 0 0 0 0 0 0 0 ...
#>   $ gear: num   4 4 4 3 3 3 3 4 4 4 ...
#>   $ carb: num   4 4 1 1 2 1 4 2 2 4 ...
```

This shows that the mtcars variable is a data frame with 11 variables (columns) and 32 rows (observations). We can easily convert the data frame to a list using the function as.list(). When processing data frames with lapply(), the most important thing to remember is that the data frame will be processed column-by-column. This can be explored in the following example, which takes the first three columns of mtcars and then calculates the average value for each of these columns. Note we also make use of the function subset() and the R pipe operator |> in order to create a small data processing workflow.

```
s1 <- mtcars |>
         subset(select=c("mpg","cyl","disp")) |>
         lapply(function(x)mean(x))
s1
#> $mpg
#> [1] 20.09
#>
#> $cyl
#> [1] 6.188
#>
#> $disp
#> [1] 230.7
```

### 5.6.3  Adding extra arguments to apply() and lapply()

There may be situations where a processing task requires more information to be passed to the function, and for this, additional arguments can be passed into the functions lapply() and apply(). These optional arguments are listed after the function is defined, and here we show an example of how additional arguments, a and b, are added and used as part of a standard function call.

```
m2 <- matrix(1:4,nrow = 2)
m2
#>      [,1] [,2]
#> [1,]    1    3
#> [2,]    2    4
apply(m2,1,function(x,a,b){a*x+b},b=10,a=2)
#>      [,1] [,2]
#> [1,]   12   14
#> [2,]   16   18
```

## 5.7 Mini-case 1: Modelling social networks using matrices

The first mini-case involves the study of networks, which are at the heart of some of the most revolutionary technologies of the 21st century, and impact many areas, including science, technology, business, and nature (Barabási, 2016). In this example, we can model connections between users in a simulated social media network, where each user can follow other users, and can also be followed by users (but cannot follow themselves). Our task is as follows:

- First, we generate an adjacency matrix $A$ which allows us to keep track of the links between $N$ users. The matrix $A$ has $N$ rows and $N$ columns (i.e., it is a square matrix), where $A_{ij} = 1$ if $person_i$ follows $person_j$, and $A_{ij} = 0$ if $person_i$ does not follow $person_j$. The first step is to create 100 random numbers, with the arbitrary probability that 20% of these have the value 1. A square matrix is then created, and the rows and columns are named P-1 through to P-10. To complete the logic of the adjacency matrix, all the diagonal elements are set to zero using the R function diag(), given that people cannot follow themselves.

```
# Set the number of users
N = 7
# Set the seed to ensure the matrix can be reproduced
set.seed(100)
# Sample, with replacement, 100 numbers, with 20% of a 1
v <- sample(0:1,N^2,replace = T,prob = c(0.6,.4))
# Create the matrix of 10x10
m <  matrix(v,nrow=N)

# Set the row and column names
rownames(m) <- paste0("P-",1:N)
colnames(m) <- rownames(m)

# Set the diagonal to zero, and display
diag(m) <- 0
m
#>      P-1 P-2 P-3 P-4 P-5 P-6 P-7
#> P-1    0   0   1   1   0   1   1
#> P-2    0   0   1   0   0   0   1
#> P-3    0   0   0   1   0   1   1
#> P-4    0   1   0   0   1   1   0
#> P-5    0   1   0   0   0   0   1
#> P-6    0   0   1   1   1   0   1
#> P-7    1   0   0   1   1   1   0
```

- Second, based on the CRAN package `igraph`, we can visualize the adjacency matrix as a network graph. Once the library is installed, sending the matrix to the `igraph` function `graph_from_adjacency_matrix()` will return a list of connections, which are then visualized using `plot()`. This network output generated by `igraph` is shown in Figure 5.1.

```
# Include the igraph library
library(igraph)
```

We then create `g1` from the matrix `m`.

```
# Create the igraph data structure from m and display
g1 <- graph_from_adjacency_matrix(m)
g1
#> IGRAPH bc44e6f DN-- 7 22 --
#> + attr: name (v/c)
#> + edges from bc44e6f (vertex names):
#>  [1] P-1->P-3 P-1->P-4 P-1->P-6 P-1->P-7 P-2->P-3 P-2->P-7
#>  [7] P-3->P-4 P-3->P-6 P-3->P-7 P-4->P-2 P-4->P-5 P-4->P-6
#> [13] P-5->P-2 P-5->P-7 P-6->P-3 P-6->P-4 P-6->P-5 P-6->P-7
#> [19] P-7->P-1 P-7->P-4 P-7->P-5 P-7->P-6

# Plot the graph (commented out here)
# plot(g1,vertex.size=20,vertex.color="red")
```

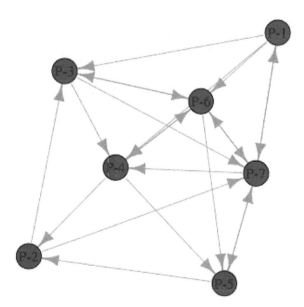

**FIGURE 5.1** Network visualization of the adjacency matrix

- Third, we can update the matrix to sum all of the rows (which will show the number of people a person follows), and the sum of all columns (showing the number of followers a person has). This can be completed using the apply() function described earlier. The name of the person with the maximum for each can also be retrieved using the which() function.

```
row_sum <- apply(m,1,sum)
row_sum
#> P-1 P-2 P-3 P-4 P-5 P-6 P-7
#>   4   2   3   3   2   4   4
names(which(row_sum==max(row_sum)))
#> [1] "P-1" "P-6" "P-7"


col_sum <- apply(m,2,sum)
col_sum
#> P-1 P-2 P-3 P-4 P-5 P-6 P-7
#>   1   2   3   4   3   4   5
names(which(col_sum==max(col_sum)))
#> [1] "P-7"
```

- Finally, we can update the matrix to include the additional information using the cbind() function.

```
m1 <- cbind(m,TotFollowing=row_sum)
m1
#>      P-1 P-2 P-3 P-4 P-5 P-6 P-7 TotFollowing
#> P-1    0   0   1   1   0   1   1            4
#> P-2    0   0   1   0   0   0   1            2
#> P-3    0   0   0   1   0   1   1            3
#> P-4    0   1   0   0   1   1   0            3
#> P-5    0   1   0   0   0   0   1            2
#> P-6    0   0   1   1   1   0   1            4
#> P-7    1   0   0   1   1   1   0            4
```

## 5.8   Mini-case 2: Creating a pipeline for processing data frames

For the data frame mtcars, the following processing actions will be taken:

- Two columns from mtcars will be selected, mpg and disp.
- A new column kpg will be added, which converts mpg to kilometers per gallon, using the multiplier 1.6.

- A new column `dm_ratio` will be added, which is the ratio of `disp` and `mpg`.
- The first six observations will then be shown.

The solution uses the R pipe operator, which chains the sequence of data processing stages together. The functions `subset()` and `transform()` are used, the first to select a subset of columns, the second to add new columns to the data frame.

```
mtcars_1 <- mtcars |>                          # the original data frame
            subset(select=c("mpg","disp")) |> # select 2 columns
            transform(kpg=mpg*1.6,             # Add first column
                      dm_ratio=disp/mpg) |>    # Add second column
            head()                             # Subset 1st 6 records
mtcars_1
#>                       mpg disp   kpg dm_ratio
#> Mazda RX4            21.0  160 33.60    7.619
#> Mazda RX4 Wag       21.0  160 33.60    7.619
#> Datsun 710          22.8  108 36.48    4.737
#> Hornet 4 Drive      21.4  258 34.24   12.056
#> Hornet Sportabout   18.7  360 29.92   19.251
#> Valiant             18.1  225 28.96   12.431
```

While it is a simple example, it does show the utility of being able to take a rectangular data structure (the data frame) as input, and then apply a chain of operations to this in order to arrive at a desired output, which is then stored in the variable `mtcars_1`.

## 5.9  Summary of R functions from Chapter 5

The following are functions that can be used to process a matrix and a data frame.

| Function | Description |
|---|---|
| as.data.frame() | Converts a tibble to a data frame |
| apply() | Iterates over rectangular data, by row or by column. |
| cbind() | Adds a new vector as a matrix column. |
| colnames() | Set (or view) the column names of a matrix. |
| colMeans() | Calculates the mean of each column in a matrix. |
| colSums() | Calculates the sum of each column in a matrix. |
| data.frame() | Constructs a data frame. |
| diag() | Sets a matrix diagonal, or generates an identity matrix |
| dim() | Returns (or sets) the matrix dimensions. |
| dimnames() | Returns the row and column names of a matrix. |
| eigen() | Calculates matrix eigenvalues and eigenvectors. |
| factor() | Encode a vector as a factor. |
| is.matrix() | Checks to see if the object is a matrix. |
| matrix() | Creates a matrix from the given set of arguments. |
| rbind() | Adds a vector as a row to a matrix. |
| rownames() | Sets (or views) the row names of a matrix. |
| t() | Returns the matrix transpose. |
| rowMeans() | Calculates the mean of each matrix row. |
| rowSums() | Calculates the sum of each matrix row. |
| subset() | Subsets data frames which meet specified conditions. |
| tibble() | Constructs a tibble (tibble package). |
| as_tibble() | Converts a data frame to a tibble (tibble package). |
| transform() | Add columns to a data frame. |

## 5.10   Exercises

1.  Use the following initial code to generate the matrix `res`.

```
set.seed(100)
N=10
CX101 <- rnorm(N,45,8)
CX102 <- rnorm(N,65,8)
CX103 <- rnorm(N,85,25)
CX104 <- rnorm(N,60,15)
CX105 <- rnorm(N,55,15)

res
#>           CX101 CX102  CX103 CX104 CX105
#> Student-1 40.98 65.72  74.05 58.63 53.48
#> Student-2 46.05 65.77 104.10 86.36 76.05
#> Student-3 44.37 63.39  91.55 57.93 28.35
```

```
#> Student-4   52.09 70.92 104.34 58.33 64.34
#> Student-5   45.94 65.99  64.64 49.65 47.17
#> Student-6   47.55 64.77  74.04 56.67 74.83
#> Student-7   40.35 61.89  66.99 62.74 49.55
#> Student-8   50.72 69.09  90.77 66.26 74.79
#> Student-9   38.40 57.69  56.06 75.98 55.66
#> Student-10  42.12 83.48  91.18 74.55 26.82
```

2. The matrix `res` (from the previous question) has values that are out of the valid range for grades (i.e., greater than 100). To address this, all out-of-range values should be replaced by NA. Use `apply()` to generate the following modified matrix.

`res_clean`
```
#>             CX101 CX102 CX103 CX104 CX105
#> Student-1   40.98 65.72 74.05 58.63 53.48
#> Student-2   46.05 65.77    NA 86.36 76.05
#> Student-3   44.37 63.39 91.55 57.93 28.35
#> Student-4   52.09 70.92    NA 58.33 64.34
#> Student-5   45.94 65.99 64.64 49.65 47.17
#> Student-6   47.55 64.77 74.04 56.67 74.83
#> Student-7   40.35 61.89 66.99 62.74 49.55
#> Student-8   50.72 69.09 90.77 66.26 74.79
#> Student-9   38.40 57.69 56.06 75.98 55.66
#> Student-10  42.12 83.48 91.18 74.55 26.82
```

3. The matrix `res_clean` (from the previous question) has NA values, and as a work around, it has been decided to replace these values with the average subject mark. Write the code (using `apply()`) to generate the matrix `res_update`.

`res_update`
```
#>             CX101 CX102 CX103 CX104 CX105
#> Student-1   40.98 65.72 74.05 58.63 53.48
#> Student-2   46.05 65.77 76.16 86.36 76.05
#> Student-3   44.37 63.39 91.55 57.93 28.35
#> Student-4   52.09 70.92 76.16 58.33 64.34
#> Student-5   45.94 65.99 64.64 49.65 47.17
#> Student-6   47.55 64.77 74.04 56.67 74.83
#> Student-7   40.35 61.89 66.99 62.74 49.55
#> Student-8   50.72 69.09 90.77 66.26 74.79
#> Student-9   38.40 57.69 56.06 75.98 55.66
#> Student-10  42.12 83.48 91.18 74.55 26.82
```

4. Use the `subset()` function to generate the following tibbles from the tibble `ggplot2::mpg`. Use the R pipe operator (|>) where necessary.

```
# The car with the maximum displacement, with a subset of features
max_displ
#> # A tibble: 1 x 6
#>   manufacturer model      year disrl    cty cluss
#>   <chr>        <chr>     <int> <dbl> <int> <chr>
#> 1 chevrolet    corvette   2008     7    15 2seater

# All 2seater cars, with selected columns
two_seater
#> # A tibble: 5 x 6
#>   class   manufacturer model    displ year   hwy
#>   <chr>   <chr>        <chr>    <dbl> <int> <int>
#> 1 2seater chevrolet    corvette   5.7  1999    26
#> 2 2seater chevrolet    corvette   5.7  1999    23
#> 3 2seater chevrolet    corvette   6.2  2008    26
#> 4 2seater chevrolet    corvette   6.2  2008    25
#> 5 2seater chevrolet    corvette     7  2008    24

# The first six audi cars, with selected columns
# Also, implement the solution without the subset() function
audi_6
#> # A tibble: 6 x 5
#>    year manufacturer model displ   cty
#>   <int> <chr>        <chr> <dbl> <int>
#> 1  1999 audi         a4      1.8    18
#> 2  1999 audi         a4      1.8    21
#> 3  2008 uudi         a4      2      20
#> 4  2008 audi         a4      2      21
#> 5  1999 audi         a4      2.8    16
#> 6  1999 audi         a4      2.8    18
```

5. Generate the following network using matrices and `igraph`. All nodes
   are connected to each other (i.e., it is a fully connected network).

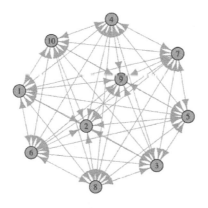

# 6

## The S3 Object System in R

Everything that exists in R is an object. Everything that happens
in R is a function call.

— John Chambers (Chambers, 2017)

## 6.1 Introduction

While the material in this chapter is more challenging, it is definitely worth the
investment of your time, as it should provide you with additional knowledge
of how R works. It highlights the importance of the S3 object oriented system,
which is ubiquitous in R. An important idea underlying S3 is that we want
to make things as easy as possible for users, when they call functions and
manipulate data. S3 is based on the idea of coding *generic functions* (for
example, the base R function summary() is a generic function), and then writing
specific functions (often called methods) that are different implementations of
a generic function. The advantage of the S3 approach is that it shields a lot of
complexity from the user, and makes it easier to focus on the data processing
workflow.

Upon completing the chapter, you should understand:

- Why calls by different objects to the same generic function can yield different
  results.
- The difference between a base object and an object oriented S3 object.
- The importance of attributes in R, and in particular, the attributes names,
  dim, and class.
- How the class attribute is essential to the S3 object system in R, and how to
  define a class for an object.

- The role of a generic function, and the process of *method dispatch*.
- How to use an existing generic function with a newly defined class.
- How to build your own generic function.
- The mechanism of inheritance for S3, and the advantages this provides for designing software.
- Key R functions that allow you to use and understand the S3 system.
- How to solve all five test exercises.

Note there are other object-oriented systems in R that we do not cover, including S4, which is similar to S3, and reference classes (RC), which are closer to the object oriented systems found in Python and Java (Wickham, 2019).

### Chapter structure

- Section 6.2 provides an insight into how S3 works.
- Section 6.3 introduces the idea of *attributes*, which are metadata associated with an object. A number of attributes are particularly important, and they include `names`, `dim` and `class`. The `class` attribute is key to understanding how the S3 system operates.
- Section 6.4 presents the generic function object oriented approach, which is the design underlying R's S3 system. It summarizes key ideas and describes the process known as method dispatch.
- Section 6.5 shows how a developer can make use of existing generic functions, for example `print()` and `summary()`, to define targeted methods for new classes.
- Section 6.6 presents examples of how to write custom generic functions that can be used to define methods for S3 classes.
- Section 6.7 shows how inheritance can be specified in S3, and how new classes (e.g., a subclass) can inherit methods from other classes (e.g., a superclass).
- Section 6.8 shows how a simple queue system in S3 can be implemented.
- Section 6.9 provides a summary of the functions introduced in the chapter.
- Section 6.10 provides a number of short coding exercises to test your understanding of S3.

## 6.2   S3 in action

In order to get an insight into how S3 works, let's consider R's `summary()` function. You may have noticed that if you call `summary()` with two different types of data, you will get two appropriately different answers, for example:

- When summarizing results of a linear regression model, based on two variables from `mtcars`, the following information is presented. The model can be used to predict the miles per gallon (mpg) given the engine size (disp - in cubic inches), and the model coefficients are shown (the slope=−0.04122 and the intercept=29.59985). Note that at this point we are not so much interested in the results of the linear model, rather we focus on what is output when we call the function summary() with the variable mod.

```
mod <- lm(mpg~disp,data=mtcars)
summary(mod)
#>
#> Call:
#> lm(formula = mpg ~ disp, data = mtcars)
#>
#> Residuals:
#>    Min     1Q Median     3Q    Max
#> -4.892 -2.202 -0.963  1.627  7.231
#>
#> Coefficients:
#>              Estimate Std. Error t value Pr(>|t|)
#> (Intercept) 29.59985    1.22972   24.07  < 2e-16 ***
#> disp        -0.04122    0.00471   -8.75  9.4e-10 ***
#> ---
#> Signif. codes:  0 '***' 0.001 '**' 0.01 '*' 0.05 '.' 0.1 ' ' 1
#>
#> Residual standard error: 3.25 on 30 degrees of freedom
#> Multiple R-squared:  0.718,  Adjusted R-squared:  0.709
#> F-statistic: 76.5 on 1 and 30 DF,  p-value: 9.38e-10
```

- When summarizing the first six rows and two columns from the mtcars data frame, summary() provides information on each column.

```
d <- subset(mtcars,select=c("mpg","disp")) |> head()
summary(d)
#>       mpg           disp
#> Min.   :18.1   Min.   :108
#> 1st Qu.:19.3   1st Qu.:160
#> Median :21.0   Median :192
#> Mean   :20.5   Mean   :212
#> 3rd Qu.:21.3   3rd Qu.:250
#> Max.   :22.8   Max.   :360
```

Think of what's happening with these two examples. In both cases, the user calls summary(). However, depending on the variable that is passed to summary(), a different result is returned. How might this be done in R? To achieve this, R uses a mechanism known as *polymorphism*, and this is a common feature of object oriented programming languages. Polymorphism means that the same function call leads to different operations for objects of different classes (Matloff, 2011). Behind the scenes, it means that different functions (usually called *methods* in S3) are written for each individual class. So for this example, there are actually two different methods at work, one to summarize the linear model results (`summary.lm()`), and the other to summarize columns in a data frame (`summary.data.frame()`). We will now explore this mechanism in more detail.

---

## 6.3   Objects, attributes and defining S3 classes

The opening quote by John Chambers, creator of the S programming language that R is based on, is an excellent summary of R, and in particular the phrase that "everything that exists in R is an object". We can discuss this in more detail by exploring the difference in R between *base objects* and *object oriented objects*. We will first discuss base objects (also known as base types), before focusing on object oriented objects using R's S3 system. Base objects originated in S and were developed before anyone thought that S would need an object oriented system, and there are 25 different base types in R (Wickham, 2019). Common base types that we have explored to date include atomic vectors (e.g., logical, integer, double, and character), lists, functions, and environments. The function `typeof()` reveals the underlying type of an object, and we can view this again with the following examples.

```
typeof("Hello world")
#> [1] "character"
typeof(TRUE)
#> [1] "logical"
typeof(2L)
#> [1] "integer"
typeof(2.71)
#> [1] "double"
```

An important feature of R that we have not explicitly referenced to date is the idea of an *attribute*, which allows us to attach metadata to a base type such as an atomic vector or a list. The function `attr()` can be used to set an attribute, and the function `attributes()` is used to retrieve all the attributes from an object. R provides much flexibility for adding attributes to base types; for example, we can add an attribute called *test* to an atomic vector, and then retrieve that attribute.

```
x1 <- 1:5
attr(x1,"test") <- "Hello World"
attr(x1,"test")
#> [1] "Hello World"
attributes(x1)
```

```
#> $test
#> [1] "Hello World"
```

While there is flexibility to add attributes, generally, there are three special attributes that are used in R, and knowledge of these is required.

- The first special attribute (which we already have used) is **names**, which stores information on the element names in an atomic vector, list, or data frame. Previously we used the function `names()` to set these values, but because names is a special attribute, we can achieve the same result by using the `attr()` function. Usually, using `attr()` is not the best way to set vector names, and the function `names()` is used instead. But the impact of both methods is the same.

```
x2 <- 1:5
attr(x2,"names") <- LETTERS[1:5]
attributes(x2)
#> $names
#> [1] "A" "B" "C" "D" "E"
x2
#> A B C D E
#> 1 2 3 4 5
```

- The second special attribute is **dim**, and this is an integer vector that is used by R to convert a vector into a matrix or an array. Notice that setting this attribute has a similar effect of creating a matrix. Normally this mechanism is not used, and the matrix is created using the `matrix()` function,

```
# Create a vector of 4 elements
m1 <- 1:4
m1
#> [1] 1 2 3 4
# Set the "dim" attribute to 2x2 matrix
attr(m1,"dim") <- c(2,2)
m1
#>      [,1] [,2]
#> [1,]    1    3
#> [2,]    2    4
attributes(m1)
#> $dim
#> [1] 2 2
# Set the "dim" attribute to 1x4 matrix
attr(m1,"dim") <- c(1,4)
m1
#>      [,1] [,2] [,3] [,4]
#> [1,]    1    2    3    4
attributes(m1)
```

```
#> $dim
#> [1] 1 4
# Remove the attribute and revert to the original vector
attr(m1,"dim") <- NULL
m1
#> [1] 1 2 3 4
attributes(m1)
#> NULL
# Check with the usual method
m2 <- matrix(1:4,nrow=2)
m2
#>      [,1] [,2]
#> [1,]    1    3
#> [2,]    2    4
attributes(m2)
#> $dim
#> [1] 2 2
```

- The third special attribute of base types is known as **class**, and this has a key role to play in the S3 object oriented system. An S3 object is a base type *with a defined class attribute*, and we can explore this by looking at the class attribute associated with a data frame. The key point is that this new attribute *class* is defined for a data frame, and when it is defined, R views the object as an S3 object, and so can apply the mechanisms of S3 to this object. The function class() can also be used to view the class attribute.

```
# Show the class attribute using attr
attr(mtcars,"class")
#> [1] "data.frame"
# Show the class attribute using class
class(mtcars)
#> [1] "data.frame"
```

Unlike other object-oriented languages, S3 has no formal definition of a class, and to make an object an instance of a class, the class attribute is set using the class() function. This attribute must be a character string, and it is recommended to use only letters and the character "_" (Wickham, 2019). For example, we can create a base type (a list), and then make the object to be an instance of a class called "my_test".

```
v <- list(a=1,b=2,c=3)
class(v) <- "my_test"
attributes(v)
#> $names
#> [1] "a" "b" "c"
#>
```

```
#> $class
#> [1] "my_test"
```

The class attribute can also be set using the attr() function.

```
v <- list(a=1,b=2,c=3)
attr(v,"class") <- "my_test"
attributes(v)
#> $names
#> [1] "a" "b" "c"
#>
#> $class
#> [1] "my_test"
```

What is also interesting is to see how a simple manipulation of the class attribute can lead to different behaviors. First, we define a data.frame object and show its structure, attributes, and whether it is viewed as an S3 object by R. We can see that the class attribute is data.frame, and so all behaviors associated with a data frame will apply (for example, subsetting, summaries, etc.). Notice that an additional attribute row.names is used for data frames, but we do not need to use this for our examples.

```
# Create a data frame
d <- data.frame(Number=1:5,
                Letter=LETTERS[1:5],
                Flag=c(T,F,T,F,NA),
                stringsAsFactors = F)
str(d)
#> 'data.frame':    5 obs. of  3 variables:
#>  $ Number: int  1 2 3 4 5
#>  $ Letter: chr  "A" "B" "C" "D" ...
#>  $ Flag  : logi  TRUE FALSE TRUE FALSE NA

# Show its attributes
attributes(d)
#> $names
#> [1] "Number" "Letter" "Flag"
#>
#> $class
#> [1] "data.frame"
#>
#> $row.names
#> [1] 1 2 3 4 5

# Call is.object function
```

```
is.object(d)
#> [1] TRUE
```

Next we make just one *minor change* to the object d by removing the class attribute. For this, we use the function unclass(), which deletes the class attribute. This simple action totally changes the way R views the object. It is now no longer an S3 object, and instead is just viewed as its base type, which is a list. Notice that R does not stop you from doing this, and you will never get a message saying, "do you realize you have just changed the class of this object?" That's not the S3 way, as it is very flexible and relies on the programmer being aware of how S3 operates.

```
# Remove its class attribute
d <- unclass(d)
# When you show d's structure, it's no longer a data frame
str(d)
#> List of 3
#>  $ Number: int [1:5] 1 2 3 4 5
#>  $ Letter: chr [1:5] "A" "B" "C" "D" ...
#>  $ Flag  : logi [1:5] TRUE FALSE TRUE FALSE NA
#>  - attr(*, "row.names")= int [1:5] 1 2 3 4 5
# The class attribute has been removed
attributes(d)
#> $names
#> [1] "Number" "Letter" "Flag"
#>
#> $row.names
#> [1] 1 2 3 4 5
# Call is.object() function
is.object(d)
#> [1] FALSE
```

Because of this flexibility, it is best practice to write a constructor function to create the new S3 object. For example, the function new_my_test() will create an object of class "my_test".

```
new_my_test <- function(l){
  class(l) <- "my_test"
  l
}

obj <- new_my_test(list(a=1,b=2))
class(obj)
#> [1] "my_test"
```

In summary, the following points are worth remembering:

- In R, everything that exists is an object (Chambers, 2017), and objects are either base objects or object oriented (e.g., S3) objects.
- All base objects in R are of a particular type; for example, an atomic vector of characters, a function, or a list (there are 25 types in all, but in reality you don't need to be concerned with knowing all of these).
- All base objects in R can have attributes. These can be user-defined, but mostly they are special attributes defined using existing R function, for example `names`, `dim` and `class`.
- An S3 object (which we explore in detail in this chapter) must have its class attribute set to a character string in order for it "to exist" as an S3 object.
- The function `is.object()` can be used to test whether an object is an S3 object.
- Once an object is an S3 object, you can perform many interesting things with it.

So what's the benefit of doing this, and where do we go from here?

- First, you can write specific implementations of available generic functions, for example, writing a version of `summary()` or `plot()` for your own list objects.

- Second, you can also define your own generic functions, and add these to your codebase in order to support flexible designs.

We will now explain R's generic function approach to object oriented programming.

## 6.4 The generic function approach

Earlier we introduced the concept of *polymorphism*, which is a key idea in object oriented programming. It allows a programmer to separate a function's *interface* from its *implementation*, which makes it possible to use the same function for different kinds of input (Wickham, 2019). This is known as functional object oriented programming, where S3 methods belong to generic functions. An example of a generic function is `summary()`, and two (of the many) methods that belong to this generic function include `summary.lm()` and `summary.data.frame()`. Notice that the methods that belong to the `summary()` generic function have the format *summary.class_name(object, arg1, ...)*.

Generic functions appear remarkably simple, and they typically contain just one line of code. Here is the code for the `summary()` generic function, which is part of the R system. Note that it has just one line of code, which is `UseMethod("summary")`.

```
summary
#> function (object, ...)
#> UseMethod("summary")
#> <bytecode: 0x13190f5f0>
#> <environment: namespace:base>
```

However, behind the scenes, the R system does a lot of work to find a way to implement this function, and here are the main steps.

- When the user calls a generic function, they pass an object, which is usually has S3 class attribute. For example, `summary(mod)`, which we explored earlier, calls `summary()` (a generic function) with the object `mod`, which is a list that is also an S3 class with the class attribute set to "lm".

- R will then invoke *method dispatch* (Wickham, 2019), which aims to find the correct method to call. Therefore, the only line of code needed in the generic function is a call to `UseMethod(generic_function_name)`, and, in this case, R will find the class attribute (e.g., "lm"), and find the method (e.g., `summary.lm()`).

- Once R finds this method, R will call the method with the arguments, and let the method perform the required processing.

- The object returned by the method will then be routed back to the original call by the generic function.

We can see how this works by writing our own test (and not very useful!) version of the method `summary.lm()` in our global workspace, so that it will be called before R's function `summary.lm()`. This new method returns the string "Hello World". Note that we use the "method" terminology because it indicates that the function is associated with a generic function. In reality, the mechanism of writing a method is the same as writing a function.

```
# Redefining summary.lm() to add in the global environment
summary.lm <- function(o)
{
  "Hello world!"
}
# Show the output from the new version of summary.lm
summary(mod)
#> [1] "Hello world!"
# Remove the function from the global environment
rm(summary.lm)
```

In general, the R function `methods()` is helpful, as it will return a list of available methods for a given generic function, where the input is a string version of the method. We can see this in the following call, which lists all the different methods associated with the generic function `summary()`.

```
m_list <- methods("summary")
length(m_list)
#> [1] 45
m_list[1:5]
#> [1] "summary,ANY-method"        "summary,DBIObject-method"
#> [3] "summary.aov"               "summary.aovlist"
#> [5] "summary.aspell"
which(m_list=="summary.data.frame")
#> [1] 9
which(m_list=="summary.lm")
#> [1] 19
```

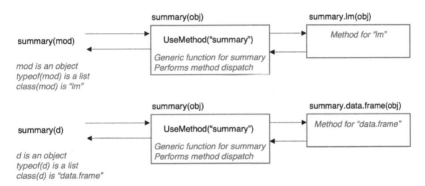

**FIGURE 6.1** An overview of polymorphism in S3

It is instructive to view the overall S3 process shown in Figure 6.1, where we revisit the earlier example. As a reminder, the two variables are mod (storing, in a list, the results of a linear regression model) and d (storing the results following subsetting of a data frame). The diagram shows what happens when the generic function summary() is called for each object.

The generic function summary() takes the input, and then performs *method dispatch*, which is the act of finding the correct method to call, based on the variable's class. In doing this, for the first example, method dispatch finds the method summary.lm(). For the second example, method dispatch calls summary.data.frame().

And this captures the elegant idea of separating the interface (i.e., the generic function) from the implementation (i.e., the method). It has a huge benefit for users of R in that they don't have to worry about finding the specific method, they simply call the generic function with the S3 class and let the S3 system in R do the rest!

## 6.5    Using an existing generic function

When using polymorphism with S3, most likely you will use existing generic
functions already available within R, for example, summary(), plot(), and
print(). Here, we will show an example of this process and focus on the
generic function print(), which formats the output from an object when it
is typed in the console. Before looking at the first of two examples, we can
confirm that print() is indeed a generic function, with the giveaway feature
that contains just one line of code, and that this line of code invokes the
function UseMethod().

```
print
#> function (x, ...)
#> UseMethod("print")
#> <bytecode: 0x1370619a0>
#> <environment: namespace:base>
```

We will look at building our own print method for a test class, where the
test class is set as an attribute on an atomic vector. The class is called
"my_new_class", and therefore the method, associated with the generic function print is named print.my_new_class(). Inside this method, we use the
cat() function to print a simple message. "Hello World!". Therefore, if the
S3 class for a variable is "my_new_class", the method print.my_new_class()
will be selected by R using method dispatch. Note that this is not a useful
function, the goal is just to indicate how to write a new method that is linked
to the generic function print().

```
# define an atomic vector
v <- c(10.1, 31.2)
# Show v at the console
v
#> [1] 10.1 31.2
# Add a class attribute
class(v) <- "my_new_class"

# Implement a print method for this class, and load.
print.my_new_class <- function(o){
  cat("Hello World!\n")
}
# Show v again at the console
v
#> Hello World!
```

A number of points are worth highlighting when viewing this code:

- When v is first printed, it just shows the regular view of an atomic vector.
- With the assignment of the class attribute ("my_new_class"), and the writing of the print method for the class, and future printing of v will now call this new function (print.my_new_class()), and therefore the string "Hello World!" is printed.
- To revert to the default printing of v, the function unclass(v) will remove the class attribute, and this will mean that the method print.my_new_class() will no longer be called.
- Furthermore, R is so flexible when setting class attributes, and it will not stop you from doing what seem like unusual things. For example, our earlier variable mod, whose class is "lm" could have its class attribute reassigned to "my_new_class", and this will print "Hello World!" when the object is printed in the console.

```
class(mod) <- "my_new_class"
mod
#> Hello World!
```

While this example showed a simple atomic vector, it's more likely that you will be using S3 classes with a list data structure and then generate new methods from some of R's generic functions. However, in addition to using available generic functions, R does provide the facility to write your own generic functions.

## 6.6 Custom built generic functions

There may be cases when a developer needs to create their own generic function, and associated methods. Here we explore a simple example, where we can create objects (from the class "account"), with the following features:

- Each object has a list structure that contains two variables, the account number and the balance.
- A constructor is used to create a new account object.
- The "account" class has two new generic functions: debit(), which reduces the balance by a given amount, credit(), which increases the balance.
- There are three methods defined as debit.account(), credit.account() and print.account()

First, we define the constructor. This defines the data (in a list), and it also specifies the class ("account").

```
# Define a constructor to create the object
new_account <- function(num,op_bal){
```

```
    l <- list(number=num,balance=op_bal)
    class(l) <- "account"
    l
}
```

Next we define the two new generic functions. Note that each generic function has just *one line of code* - UseMethod() - and we have added an extra argument to the function, as we need to pass a value for the debit and credit amount into each of these functions. First, we define the debit() generic function.

```
# Define a generic function to debit an account
debit <- function(o,amt){
    UseMethod("debit")
}
```

Next, we define the credit() generic function.

```
# Define a generic function to credit the account
credit <- function(o,amt){
    UseMethod("credit")
}
```

With the generic functions defined, we now define three new methods for the class "account". The first is the debit.account() method. When the debit() generic function is called with an object whose class is "account", R will dispatch the call to debit.account().

```
# Create a debit method
debit.account <- function(o,amt){
    o$balance <- o$balance - amt
    o
}
```

The second is the credit.account() method. When the credit() generic function is called with an object whose class is "account", R will dispatch the call to credit.account().

```
# Create a credit method
credit.account <- function(o,amt){
    o$balance <- o$balance + amt
    o
}
```

Finally, we add a method for printing the object. When the print() generic function is called with an object whose class is "account", R will dispatch the call to print.account().

```
# Create a print method
print.account <- function(o){
  cat("Acc# =",o$number,"  Balance =",o$balance,"\n")
}
```

To test the code, we first create an object a1. Note that the type of this object is a list and the class is "account". Recall that when we type a1 at the console, R will actually call the method print.account().

```
# Create an object
a1 <- new_account("1111",200)
typeof(a1)
#> [1] "list"
class(a1)
#> [1] "account"
a1
#> Acc# = 1111   Balance = 200
```

We then call the debit function to remove money from the account. Note that the object a1 must be updated, namely, it receives an updated copy of the list values.

```
# Debit the account
a1 <- debit(a1, 100)
a1
#> Acc# = 1111   Balance = 100
```

Finally, we credit the account through calling the generic function credit(), and viewing the updated value of a1.

```
# Credit the account
a1 <- credit(a1, 300)
a1
#> Acc# = 1111   Balance = 400
```

In summary, this section has shown how we can create our own generic functions: however, even when we create these, notice that we still use one of R's available generic functions print(). This tends to be the pattern of using S3; in many cases we can add methods that use existing generic functions, and on occasion we may decide to create our own generic functions.

## 6.7 Inheritance with S3

In object oriented languages, the idea of inheritance is to create new classes as specialized version of older ones (Matloff, 2011), and S3 provides this mechanism. The advantage of using inheritance is that the existing class methods of a *superclass* can be used by *subclass*. The relationship between a subclass and a superclass can be viewed as an *is-a* type of relationship, and this idea is widely used in object oriented systems. For example, if a *bank account* was a superclass, then a *savings account* could be a subclass.

To explore how inheritance works in S3, we first define what will be a *superclass*, which is the class "student". This has a number of variables that describe a student and that are stored in a list (ID, name, and undergraduate degree), and the class attribute is set to "student". We also associate a method with the student class (`print.student()`), and show the output. Note that in this simple example, we have not used a constructor function to create the s1 object.

```
s1 <- list(ID="S001",
           Name="J. Jones",
           UGDegree="Economics")
class(s1) <- "student"


print.student <- function(o){
  cat("ID = ",o$ID," Name = ",o$Name," UGDegree=",o$UGDegree,"\n")
}
s1
#> ID =   S001   Name =   J. Jones   UGDegree= Economics
```

Next, we define our *sublcass*, where the class is *postgrad*. This class has one extra variable for this type of student, which is their current postgrad degree. We define the list object that contains the data for this new student, and this data is stored in the variable s2.

```
s2 <- list(ID="S001",
           Name="A. Smith",
           UGDegree="Maths",
           PGDegree="Operations Research")
```

We now must establish the relationship between the two classes, and this is performed by using two string values in the class attribute, and S3 will interpret this (from left to right) as the subclass to superclass relationship. In this case, it defines that *"postgrad" is a subclass of "student"*, and both of these are now S3 classes.

```
class(s2) <- c("postgrad","student")
class(s2)
#> [1] "postgrad" "student"
```

Given that the relationship is established, we can now see the benefit of this structure. For example, all methods associated with the class "student" are now available for the class "postgrad". We can see this by calling the generic function `print(s2)` and observing the result.

```
print(s2)
#> ID =  S001  Name =  A. Smith  UGDegree= Maths
```

What has happened here is that the method print.student() has been called. This is controlled by the method dispatch function in the generic function print(). Because the class attribute on the s2 object is a vector with two values, it takes the first value "postgrad" and searches for the function print.postgrad().

Because that function does not yet exist, it then moves to the superclass "student" and searches for the function print.student(). As that function does exist, it then is called by the generic function. Therefore, the method dispatch process has a key role to play. It will navigate the class hierarchy, starting at the base class, and continuing through to the superclasses, until it finds a matching function.

We can now write a method for the subclass "postgrad", and this is called print.postgrad().

```
print.postgrad <- function(o){
    cat("This is a postgraduate student, PG course = ",o$PGDegree,"\n")
    cat("Additional information on student...\n")
    class(o) <- class(o)[-1]
    print(o)
}
s2
#> This is a postgraduate student, PG course =  Operations Research
#> Additional information on student...
#> ID =  S001  Name =  A. Smith  UGDegree= Maths
```

A number of aspects are worth noting within the print.postgrad() method:

- The first line prints the information that is unique to the class *postgrad*, which is the list element accessed by o$PGDegree.

- The remaining information can be accessed via a call to the superclass method. To do this, and to ensure that the generic function will perform the method dispatch, we simply have to redefine the object o inside the function. This is done by removing the first element (which is "postgrad") from the class attribute. The consequence of this is that o is now an object of class "student".

- The call to the generic function print() will then result in a subsequent call to the method print.student(), and so the superclass method will be invoked, and all the remaining list elements of s2 will be displayed.

As we move through the textbook, we will see examples of inheritance in classes that we will use. The relationship between tibble and a data.frame is an example worth exploring. Below we show the class for mtcars, which is data.frame S3 class. Next we highlight the class for `ggplot2::mpg` and we can see the inheritance hierarchy where the base class is "tbl_df", its superclass is "tbl" and the next class is the hierarchy is "data.frame". This confirms that tibbles are extensions of data frames.

```
class(mtcars)
#> [1] "data.frame"
```

```
class(ggplot2::mpg)
#> [1] "tbl_df"      "tbl"           "data.frame"
```

In summary, it is important to understand that S3 supports inheritance, and this provides advantages when you are designing your own S3 classes. For example, you could take an existing class in R, and add a subclass to this in order to reuse available code. Or, when designing your own set of classes, you can make use of inheritance to capture possible *is-a* relationships, and so maximize code reuse within your solution.

## 6.8 Mini-case: Creating a queue S3 object

The aim of this mini-case is to show how to create a queue object for a company's production line (first-in first-out), using R's S3 object system. This example will have the following features:

- It will contain a constructor function `new_queue()` that creates a list object with queue attributes, and also ensures that this is an S3 object of the class "queue". The queue attributes will be *queue_name, queue_description, waiting*, a list of products waiting to be processed, and *completed*, a list of products that have been completed.
- It will create a new S3 method from the generic function `print()`, and this will summarize the overall queue details.
- It will create two new generic functions, `add()`, which adds an element to the queue, and `process()`, which removes an element from the queue and adds it to this vector *completed*.
- To simplify the example, character atomic vectors will be used to store (1) the products in the queue and (2) the products that have been processed.

```
new_queue <- function(name,description){
  l <- list(queue_name=name,
            queue_description=description,
            waiting=vector(mode="character"),
            completed=vector(mode="character"))
  class(l) <- "queue"
  l
```

```
}

q1 <- new_queue("Q1","Products queue")
str(q1)
#> List of 4
#>  $ queue_name       : chr "Q1"
#>  $ queue_description: chr "Products queue"
#>  $ waiting          : chr(0)
#>  $ completed        : chr(0)
#>  - attr(*, "class")= chr "queue"
class(q1)
#> [1] "queue"
```

The object q1 is a list with four elements, with the class attribute set to "queue". We can now add a new function and utilize the generic function capability of S3 and implement our version of the function print().

```
print.queue <- function(q){
  cat("Queue name = ",q$queue_name,"\n")
  cat("Queue Description = ",q$queue_description,"\n")
  cat("Waiting <",length(q$waiting),"> Products Waiting = ",
      q$waiting,"\n")
  cat("Processed <",length(q$processed),"> Products Processed = ",
      q$processed,"\n")
}

q1
#> Queue name =  Q1
#> Queue Description =  Products queue
#> Waiting < 0 > Products Waiting =
#> Processed < 0 > Products Processed =
```

The new function print.queue() will now print this specific queue information when it is called by the generic function print(). The generic function print() has the following structure, with just one line of code that contains the call UseMethod("print).

```
print
#> function (x, ...)
#> UseMethod("print")
#> <bytecode: 0x1370619a0>
#> <environment: namespace:base>
```

And, interestingly, this call by the generic function to print.queue is removed once you delete the class attribute for q1. Instead, the data is shown as a list.

```
unclass(q1)
#> $queue_name
#> [1] "Q1"
#>
#> $queue_description
#> [1] "Products queue"
#>
#> $waiting
#> character(0)
#>
#> $completed
#> character(0)
```

In order to complete the example, and given that there are no generic functions named add() and process(), we need to create two generic functions. Similar to all generic functions in S3, they both contain one line of code, and the argument passed to UseMethod is the generic function name. Note that the function can take as many arguments as required, and these in turn will be forwarded to the method by the S3 system.

```
add <- function(q,p){
  UseMethod("add")
}

process <- function(q){
  UseMethod("process")
}
```

Finally, we can add the two specific functions for the "queue" class, and these will be invoked by the generic functions in order to perform the work. The first of these is the add.queue() function. For this we simply add the new queue object to the existing queue.

```
add.queue <- function (q, p) {
  q$waiting <- c(q$waiting, p)
  q
}
```

Notice that we must return the updated list from this function, in order to have the most recent copy of the list. The function is now available for use, and can be tested by adding a product to the queue ("P-01") and then printing the updated object.

```
q1 <- add(q1,"P-01")
q1
#> Queue name =   Q1
#> Queue Description =   Products queue
```

```
#> Waiting < 1 > Products Waiting =   P-01
#> Processed < 0 > Products Processed =
```

Our final task is to implement a version of the generic function process() that will take the oldest item in the queue, remove it, and place it in the processed vector. This code requires an important check: if the queue is empty, then nothing should happen, and the queue object is returned, unchanged. If there is an item on the queue, it is (1) added to the processed element, and (2) removed from the queue by using the subsetting mechanism q\$waiting[-1], which removes the first element.

```
process.queue <- function(q){
  if(length(q$waiting) == 0){
    cat("Cannot process queue as it is empty!\n")
    return(q)
  }
  q$processed <- c(q$processed, q$waiting[1])
  q$waiting    <- q$waiting[-1]
  q
}
```

We can test this new function as follows.

```
q1
#> Queue name =   Q1
#> Queue Description =   Products queue
#> Waiting < 1 > Products Waiting =   P-01
#> Processed < 0 > Products Processed =
q1 <- process(q1)
q1
#> Queue name =   Q1
#> Queue Description =   Products queue
#> Waiting < 0 > Products Waiting =
#> Processed < 1 > Products Processed =   P-01
```

In summary, this small example shows the utility of implementing S3 along with a list structure. The benefits for the programmer using this type of design are:

- It's easy to create the new object by calling the constructor function new_queue(). Note that the user of the queue does not need to know that a "queue" object has been created, although they can see this by examining the class attribute through the function class().
- Tailored information is provided on the object when it is printed, and this is because we created a new function called print.queue(), which is automatically called by the generic function print().

- Adding elements to the queue is straightforward. The function add() is called, with two parameters, the queue and the product to be added. Notice that the user never needs to know that, behind the scenes, the new function add.queue() is actually doing all the work. This is a major benefit of S3, as the implementation method is hidden from the user.
- Processing the queue is also simplified, the generic function process() is called, and this relays the call to process.queue(), and the updated queue object is returned.

To summarize, every time we use R, we are almost certainly invoking the S3 system. The chapter has shown how knowledge of the S3 can enhance our understanding of how R works and also provide a software design approach that can enhance the flexibility of the functions we design, and so provide the users of our software with efficient and flexible workflows.

## 6.9   Summary of R functions from Chapter 6

The following functions can be used as part of S3 programming in R.

| Function | Description |
| --- | --- |
| attr() | Get or set object attributes. |
| attributes() | Return object attributes. |
| class() | Return object S3 class. |
| is.object() | Checks to see if an object has a class attribute. |
| unclass() | Removes the class attribute of an object. |
| methods() | Lists all available methods for a generic function. |
| UseMethod() | Method to invoke method dispatch for S3. |
| glm() | Used to fit generalized linear models. |

## 6.10   Exercises

1. Show how you can generate the following object v, which has a number of attributes.

```
v
#> A B  C D E
#> 1 3  5 7 9
#> attr(,"class")
#> [1] "test_class"
```

Create a new `summary.class_name()` method for "test_class" that displays the following.

```
summary(v)
#> Vector Size =  5  Median Value = 6
```

2. Write a new print method for an "lm" class (this will reside in the global environment) that will simply just display the model coefficients. Use the R function `coef` to display the coefficients as a vector. Here is sample output from the new print function.

```
mod1 <- lm(mpg~disp,data=mtcars)
summary(mod1)
#> Coefficient names = [ (Intercept) disp ]
#> Coefficient values = [ 29.6 -0.04122 ]
```

3. Write a simple S3 object for managing inventory levels. Inventory, known as a stock keeping unit (sku), should have an ID (character), the number of units on hand, and a history of transactions in sequence (e.g., <ADD 100> <REMOVE 50>, etc.), where the most recent transaction is added at the end of this vector. The class name should be "sku". Three methods should be implemented: add(), remove(), and print(), and two new generic functions should be written. A constructor is required to create an inventory object. Your solution should replicate the following output.

```
# Create an sku object
sku1 <- new_sku("SKU_001",onhand = 100)
sku1
#> SKU = SKU_001  OnHand= 100
#> #TX =  0  TX =
sku1 <- add(sku1,200)
sku1
#> SKU = SKU_001  OnHand= 300
#> #TX =  1  TX = <ADD 200>
sku1 <- remove(sku1,299)
sku1
#> SKU = SKU_001  OnHand= 1
#> #TX =  2  TX = <ADD 200> <REMOVE 299>
```

4. Create a linear model in R, using the call `mod2 <- lm(mpg~wt,data=mtcars)`. At the outset, you will see that the object `mod2` will have the class attribute "lm". However, you should now write a function called `modify_lm_class()` that will change the base class for `mod2` to "my_lm", while ensuring that `mod2` inherits

from the class "lm". Write a new version summary function that displays a message, before showing the usual summary output from an "lm" object. The following code should be implemented. In your method summary.my_lm(), you cannot make a direct call to the method summary.lm(); this method must be accessed via a generic function call. Think about how you might achieve this by manipulating the class attribute.

```
mod2 <- lm(mpg~wt,data=mtcars)
class(mod2)
#> [1] "lm"
mod2 <- modify_lm_class(mod2)
class(mod2)
#> [1] "my_lm" "lm"
summary(mod2)
#>
#> Welcome to my own summary of lm outputs...
#> Now here is all the information from summary.lm()...
#>
#> Call:
#> lm(formula = mpg ~ wt, data = mtcars)
#>
#> Residuals:
#>     Min      1Q Median      3Q     Max
#> -4.543 -2.365 -0.125   1.410   6.873
#>
#> Coefficients:
#>              Estimate Std. Error t value Pr(>|t|)
#> (Intercept)    37.285      1.878   19.86  < 2e-16 ***
#> wt             -5.344      0.559   -9.56  1.3e-10 ***
#> ---
#> Signif. codes:  0 '***' 0.001 '**' 0.01 '*' 0.05 '.' 0.1 ' ' 1
#>
#> Residual standard error: 3.05 on 30 degrees of freedom
#> Multiple R-squared:  0.753,   Adjusted R-squared:  0.745
#> F-statistic: 91.4 on 1 and 30 DF,  p-value: 1.29e-10
```

# Part II

# The tidyverse and Shiny

# 7

## Visualization with `ggplot2`

Data graphics provide one of the most accessible, compelling, and expressive modes to investigate and depict patterns in data.

— Benjamin S. Baumer, Daniel T. Kaplan, and Nicholas J. Horton
(Baumer et al., 2021)

## 7.1 Introduction

A core part of any data analysis and modelling process is to visualize data and explore relationships between variables. The exploratory data analysis process, which we will introduce in Chapter 12, involves iteration between selecting data, building models, and visualizing results. Within R's `tidyverse` we are fortunate to have access to a visualization library known as `ggplot2`. There are three important benefits of using this library: (1) plots can be designed in a layered manner, where additional plotting details can be added using the + operator; (2) a wide range of plots can be generated to support decision analysis, including scatterplots, histograms, and time series charts, and (3) once the analyst is familiar with the structure and syntax of `ggplot2`, charts can be developed rapidly, and this supports an iterative process of decision support.

Upon completing the chapter, you should understand the following:

- How to identify the three main elements of a plot: the data (stored in a data frame or tibble), the aesthetic function `aes()` that maps data to visual properties, and geometric objects ("geoms") that are used to visualize the data.

- How to construct a scatterplot of two variables, and add instructions to color and size points by other available variables.
- How to tailor the information relating to a graph through the lab() function; for example, setting the graph title, sub-title, and legend.
- How to create sub-plots using the functions facet_wrap() and facet_grid().
- How to use plotting functions that perform statistical transformation of raw data, for example, bar charts, histograms, and exploring co-variation using pair-wise plots.
- How themes can be used to modify plot elements, for example, changing the font size of titles.
- How to conveniently add additional lines to a plot, and also present time series data based on specific date types in R.
- How to solve all three exercises.

One point to note is that it is not feasible to comprehensively present ggplot2 in a single chapter; the aim here is to provide a solid introduction to the package, and encourage you to research the topic further in order to enhance your plots.

## Chapter structure

- Section 7.2 introduces the two main datasets used in this chapter, both contained in the package ggplot2, and are named mpg and diamonds.
- Section 7.3 shows the first example of plotting, which is the scatterplot, and we explore the relationship between two variables in a fuel economy dataset.
- Section 7.4 builds on the scatterplot example to show how additional features can be added to a plot in a convenient way, and thus layering a richer information set on the plot.
- Section 7.5 introduces a powerful feature of ggplot2 which provides a mechanism to divide a plot into a number or sub-plots.
- Section 7.6 moves beyond the scatterplot to show ggplot2 functions that transform data before plotting; examples here include bar charts and histograms.
- Section 7.7 introduces the idea of a *theme*, which provides the facility to modify the non-data elements of your plots.
- Section 7.8 shows how you can add three types of lines to your plots.
- Section 7.9 explores an addition dataset - aimsir17 - and focuses on how we can explore and format time series data.
- Section 7.10 provides a summary of the functions introduced in the chapter.
- Section 7.11 presents coding exercises to test your understanding of ggplot2.

## 7.2 Two datasets from the package `ggplot2`

Before exploring `ggplot2`, please ensure that you have downloaded the package from CRAN, using the command `install.packages("ggplot2")`. For our introduction, we will base our plots on two datasets that are already part of the `ggplot2` package. These are the tibbles `mpg` and `diamonds`. Having data stored within a tibble (or a data frame) is a prerequisite to using the many functions available in `ggplot`, and the structure of these is usually in the form of *tidy data*, where every column is a variable, and every row is an observation. Two useful and related definitions are provided for two terms (Wickham, 2016):

- A *variable* is a quantity, quality, or property that you can measure, and will have a value at the time it is measured. This value can change; for example, the temperature recorded at midnight can differ from that recorded one hour later.

- An *observation* is a set of measurements made under similar conditions (often at the same time), and an observation usually contains several values, each associated with a different variable. For example, a number of weather measurements (temperature, wind speed, etc.) can be taken at the same time and, together, comprise an observation.

In Figure 7.1, we summarize the variables in the two tibbles. The `mpg` tibble contains observations ($N = 234$) on fuel economy data from cars, and the relationship between the engine displacement (liters) and miles per gallon (both city and highway) is often explored in R textbooks. We will use this tibble to mostly explore scatterplots, as we show the relationships between variables.

ggplot2::mpg, N = 234

| Variable | Description |
|---|---|
| manufacturer | Manufacturer name |
| model | Model name |
| displ | Engine displacement (liters) |
| year | Year of manufacture |
| cyl | Number of cylinders |
| trans | Type of transmission |
| drv | Type of drive train (e.g. front wheel) |
| cty | City miles per gallon |
| hwy | Highway miles per gallon |
| fl | Fuel type |
| class | "type" of car (e.g. "compact") |

ggplot2::diamonds, N = 53,940

| Variable | Description |
|---|---|
| carat | Weight of the diamond |
| cut | Quality of the cut (categorical) |
| color | Diamond color (categorical) |
| clarity | Diamond clarity (categorical) |
| depth | Total depth percentage |
| table | Measure related to width of diamond top |
| price | Price in dollars |
| x | Length in mm |
| y | Width in mm |
| z | Depth in mm |

**FIGURE 7.1** Two tibbles used to explore ggplot2 functions

The tibble `diamonds` contains prices and other attributes of many diamonds ($N = 53{,}940$). This tibble will be used to show how `ggplot2` performs statistical transformations on data before displaying it. This includes generating bar charts for different categories of diamonds, where categories are often stored as *factors* in R, where each value is one of a wider group, for example, the `cut` of a diamond can be one of `Fair`, `Good`, `Very Good`, `Premium`, or `Ideal`.

## 7.3    Exploring relationships with a scatterplot

Creating a scatterplot to explore the relationship between two numeric variables is a convenient way to start our exploration of `ggplot2`. In general, a scatterplot requires two sets of paired values: those that are represented on the x-axis, and the corresponding values mapped to the y-axis. Before working through the examples, we load the library `ggplot2`, and also the library `tibble`, as we make use of this to create a dataset for one of our examples.

```
library(ggplot2)
library(tibble)
```

For our first example, we focus on the dataset `mpg`, and the variables *engine displacement* (in liters) represented by the variable `displ` and *city miles per gallon* recorded through the variable `cty`. Before plotting, we can view a summary of these two variables as follows. For convenience, we utilize R's pipe operator.

```
mpg |> subset(select=c("displ","cty")) |> summary()
#>      displ            cty
#>  Min.   :1.60    Min.   : 9.0
#>  1st Qu.:2.40    1st Qu.:14.0
#>  Median :3.30    Median :17.0
#>  Mean   :3.47    Mean   :16.9
#>  3rd Qu.:4.60    3rd Qu.:19.0
#>  Max.   :7.00    Max.   :35.0
```

To create the graph, the following `ggplot2` functions are called. Note that we can build a graph in *layers*, where each new layer is added to a graph using the + operator.

- First, we call `ggplot(data=mpg)` which initializes a `ggplot` object, and this call also allows us to specify the tibble that contains that data. By itself, this call actually generates an empty graph; therefore, we need to provide more information for the scatterplot.

- Next, we extend this call to include the x-axis and y-axis variables by including an addition argument (`mapping`) and the function `aes()` which describes how variables in data are mapped to the plot's visual properties. The completed command is ggplot(data=mpg,mapping=aes(x=displ,y=cty)). We can execute this command, and a more informative graph appears, as it will show the numerical range of each variable, which is based on the data in mpg. However, when you try this command, you will not see any points on the graph.

- Finally, we need to visualize the set of points on the graph, and we do this by calling the relevant *geometric object*, which is one that is designed to draw points, namely the function geom_point(). As ggplot2 is a layered system for visualization, we can use the + operator to add on new elements to a graph, and therefore the following command will generate the graph of interest, shown in Figure 7.2.

```
ggplot(data=mpg, mapping=aes(x=displ,y=cty)) +
  geom_point()
```

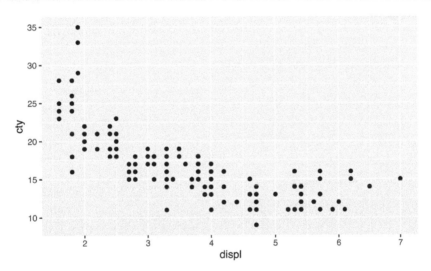

**FIGURE 7.2** Plotting displacement vs. city miles per gallon

The scatterplot is then produced showing the points. This plot can be enhanced, using the idea of *aesthetic mappings*, which we will now explore.

## 7.4   Aesthetic mappings

In the first example, we used the function aes() to specify the variables for the two axes on the plot. However, this aes() function has some nice additional features that can be used to add extra information to a plot by using data from other variables. For example, what if we also wanted to see which *class* of car each point belonged to? There are seven classes of car in the dataset, and we can see this by running the following code, where we can access a tibble's variable using the $ operator. Recall, this is made possible because the type of a tibble is a list, as we already discussed in Chapter 5.

```
unique(mpg$class)
#> [1] "compact"   "midsize"   "suv"        "2seater"
#> [5] "minivan"   "pickup"    "subcompact"
```

This information can then be used by the `aes()` function by setting the argument `color` to the particular variable we would like to color the plot by. Note the following code, and see how we have just added one additional argument to the `aes()` function, and the updated output is shown in Figure 7.3.

```
ggplot(data=mpg,mapping=aes(x=displ,y=cty,color=class))+
  geom_point()
```

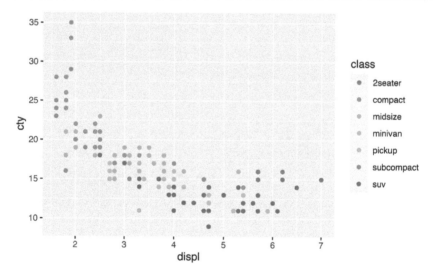

**FIGURE 7.3** Coloring the points by the class variable

With this single addition, we have added new information to the plot, where each class of car now has a different point color. The call to `ggplot()` produces a default setting for the legend, where in this case it adds it to the right side of the plot. It also labels the legend to be the same name as the variable used.

There are a number of additional arguments we can embed inside the `aes()` function, and these include `size` and `shape`, which again will change the appearance of the points, and also update the legend as appropriate. For example, we can add information to the plot that indicates the relative sizes of the number of cylinders for each observation, by setting the `size` argument.

As can be viewed in Figure 7.4, this provides information showing that cars with larger displacement values also have a higher number of cylinders.

```
ggplot(data=mpg,mapping=aes(x=displ,y=cty,color=class,size=cyl))+
  geom_point()
```

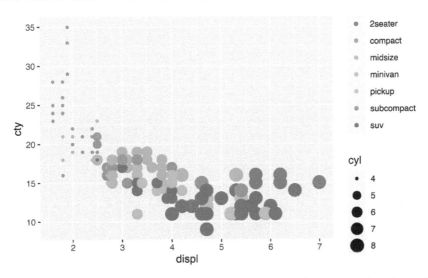

**FIGURE 7.4** Changing the size of the points to be proportional to the cylinder values

As you become familiar with the range of functions within ggplot2, you will see that the default appearance of the plot can be customized, and enhanced. A function to add further information to a plot is the `lab()`, which takes a list of name-value pairs that allows you to specify the following elements on your plot:

- `title` provides an overall title text for the plot.
- `subtitle` adds a subtitle text.
- `color` allows you to specify the legend name for the color attribute.
- `caption` inserts text on the lower right-hand side of the plot.
- `size`, where you can name the size attribute.
- `x` to name the x-axis.
- `y` to name the y-axis.
- `tag`, the text for the tag label to be displayed on the top left of the plot.

We can see the impact of using `lab()` in the updated scatterplot, where we select `displ` on the x-axis and `hwy` on the y-axis. Within the `aes()` function, we color the points by the variable `class` and size the points by the variable `cyl`. Note that we use a new approach (which is optional) where we store the result of an initial call to `ggplot()` in a variable `p1`, and then layer the labelling information onto this basic plot using the + operator. Typing p1 will then display the plot, and the revised output is displayed in Figure 7.5.

```
p1 <- ggplot(data=mpg,aes(x=displ,y=hwy,size=cyl,color=class))+
      geom_point()


p1 <- p1 +
      labs(
          title = "Exploring automobile relationships",
          subtitle = "Displacement v Highway Miles Per Gallon",
          color = "Class of Car",
          size = "Cylinder Size",
          caption = "Sample chart using the lab() function",
          tag = "Plot #1",
          x = "Displacement (Litres)",
          y = "Highway Miles Per Gallon"
      )

p1
```

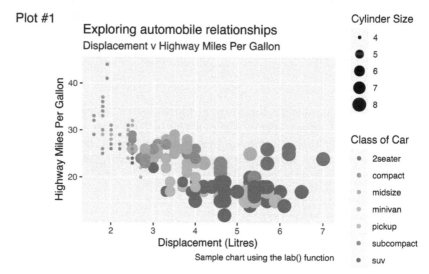

**FIGURE 7.5** Configuring the plot using the labs() function

Two further changes can be made to the appearance of a plot by altering the scales on the x-axis and the y-axis, using the functions `scale_x_continuous()` and `scale_y_continuous()`. Among the attributes that can be changed are:

- The plot limits, using the argument `limits`, where the lower and upper are specified in a vector.

- The plot tick points on each axis, using the argument `breaks`, where a vector is used to specify the points that the breaks are to be set. The function `seq()` is used, as we can specify the lower, upper, and the distance between each point.

These functions allow us to *zoom in* on a plot, and also increase the number of ticks. There is an associated warning message, as we have ignored data points outside of the new ranges; therefore, ggplot2 informs us of this. The output is display in Figure 7.6.

```
p1 <- p1 +
      scale_x_continuous(limits=c(4,5), breaks=seq(4,5,.1))+
      scale_y_continuous(limits=c(10,20),breaks=seq(10,20,1))+
      labs(caption = "This shows to zoom in on graph regions")

p1
#> Warning: Removed 192 rows containing missing values
#> (`geom_point()`).
```

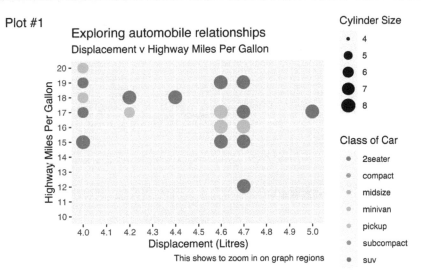

**FIGURE 7.6** Zooming in on part of a plot

## 7.5 Subplots with facets

The plots we have generated so far are a single plot that contains all the points, and this is important in order to obtain an overall sense of the relationship between the two variables. However, what if we needed to *drill down* on the plots and show, for example, the relationships for each class of car?

We have seen that there are seven car classes, and therefore, the challenge is how can we create a separate plot for each class of car. Or, in the more general case, sub-divide a plot into multiple plots based on another variable. The function `facet_wrap()` will do this in ggplot2, and all it needs as an argument is the variable for dividing the plots, which must be preceded by the tilde (~) operator. Here is the sample code to do this based on our earlier example, and the outputs is visualized in Figure 7.7. Note that the number of plots per row can be controlled by using the arguments `nrow` and/or `ncol` in the call to `facet_wrap()`.

```
ggplot(data=mpg,aes(x=displ,y=cty))+
  geom_point()+
  facet_wrap(~class)
```

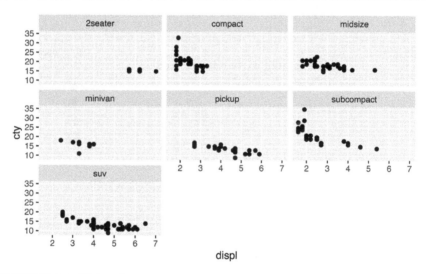

**FIGURE 7.7** Using facets to sub-divide plots

An extra variable can be added to the faceting process by using the related function `facet_grid()`, which takes two arguments, separated by the ~ operator. The first argument specifies which variable is to be mapped to each row, and the second argument identifies the variable to be represented on the columns. For example, we may want to generate 21 plots that show the type of drive (`drv`) on the columns, and the class of car (`class`) shown on each row. The following code will generate this plot, and the output is shown in 7.8.

```
ggplot(data=mpg,mapping = aes(x=displ,y=cty))+
  geom_point()+
  facet_grid(class~drv)
```

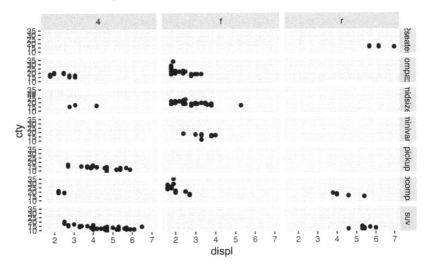

**FIGURE 7.8** Generating a grid of plots

This analysis enhances our appreciation of the `mpg` dataset, as it shows that a number of the combinations do not contain any observations; for example: front-wheel drive cars that are pickups, and rear-wheel drive compact cars. This point is illustrative: the use of faceting and sub-plots can reveal interesting patterns and relationships in the data, and their ease of use also makes them an attractive feature when undertaking exploratory data analysis.

## 7.6   Statistical transformations

The scatterplot that we have just explored takes observations from the dataset, and these points are plotted on the graph. However, there are graphs that will firstly calculate values, and then plot the results of this calculation. This is a valuable feature of `ggplot2`, because it reduces the amount of work that an analyst needs to do in order to make use of what we is termed *statistical transformations* in `ggplot2`.

Using the `diamonds` dataset that has over 50,000 observations, we will now explore four categories of these charts: (1) bar charts for summarizing frequncy data of (typically) categorical variables; (2) generating histograms and related graphs; (3) visualizing the five-number summary of a variable using a boxplot; and (4) visualizing covariation and possible relationships between sets of variables using `ggpairs()`, which is part of the `GGally` library.

## 7.6.1   Exploring count data with `geom_bar()`

In Figure 7.1 we noted that the tibble `diamonds` has three categorical variables (factors in R): `cut` (five types), `color` (seven types), and `clarity` (eight types). We can show the quantity of diamonds for each type as follows.

```
summary(diamonds$cut)
#>      Fair     Good Very Good   Premium     Ideal
#>      1610     4906    12082     13791     21551
```

These values can be visualized by calling the function `geom_bar()`, and in this case, `ggplot2` will count each type before displaying the bar chart, which is shown in Figure 7.9.

```
ggplot(data=diamonds,mapping=aes(x=cut))+
  geom_bar()
```

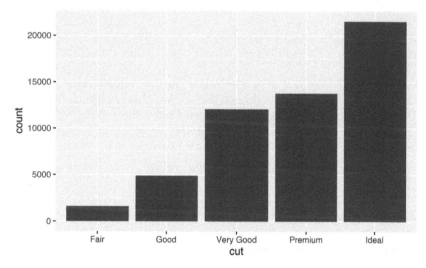

**FIGURE 7.9** A bar chart shows frequency counts of diamond cuts

A wider variety of bar charts can be constructed. For example, if we wanted to sub-divide each cut by the color, we can use the `fill` argument that is part of the `aes()` function. This provides a more informative plot, as shown in Figure 7.10.

```
ggplot(data=diamonds,mapping=aes(x=cut,fill=color))+
  geom_bar()
```

Further plots can be generated. For example, to show the colors side-by-side instead of using the default stacked format, we can set the argument `position="dodge"` as part of the `geom_bar()` function call, and this provides a comparison for the relative size of each diamond color within the different cuts. Figure 7.11 displays the output from this graph.

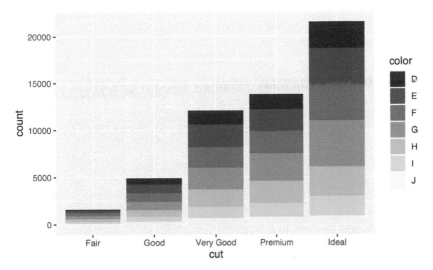

**FIGURE 7.10** Adding more information to bar chart

```
ggplot(data=diamonds,mapping=aes(x=cut,fill=color))+
  geom_bar(position="dodge")
```

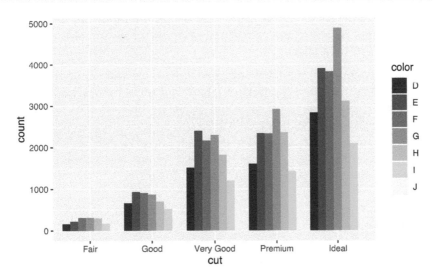

**FIGURE 7.11** Showing the chart bars side-by-side

A valuable perspective is if you wanted to see the relevant proportions of each color within a specific cut, for instance, where the column values all sum to one. This can be accomplished by setting the position argument to "fill", and the output is shown in Figure 7.12.

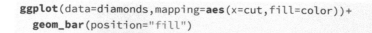

```
ggplot(data=diamonds,mapping=aes(x=cut,fill=color))+
  geom_bar(position="fill")
```

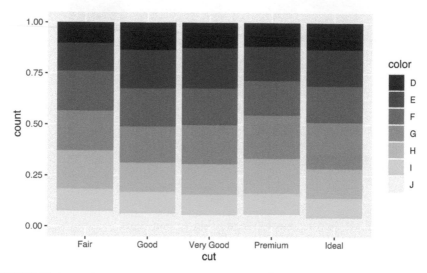

**FIGURE 7.12** Showing the overall proportions for each color

In summary, the power of geom_bar() is that it will perform the aggregations before plotting. However, what about the situation where you already had the aggregated data in a tibble?

For example, consider the following code, where we will count the number of diamonds in the clarity category (there is a more convenient way to do this using the functions group_by() and count() which are part of the dplyr package, which is covered in Chapter 8).

```
cl_cat   <- unique(as.character(diamonds$clarity))
cl_count <- unlist(lapply(cl_cat,function(x)sum(x==diamonds$clarity)))
cl_res   <- tibble(Clarity=cl_cat, Count=cl_count)
cl_res
#> # A tibble: 8 x 2
#>    Clarity Count
#>    <chr>   <int>
#> 1 SI2      9194
#> 2 SI1     13065
#> 3 VS1      8171
#> 4 VS2     12258
#> 5 VVS2     5066
#> 6 VVS1     3655
#> 7 I1        741
#> 8 IF       1790
```

In order to plot this summary data, we need to add a new argument to the geom_bar() function, which is stat="identity". This will take the raw values from the tibble, and therefore no aggregations are performed on the data. The result of this is presented in Figure 7.13.

```
ggplot(data=cl_ies,aes(x=clarity,y=Count))+
  geom_bar(stat="identity")
```

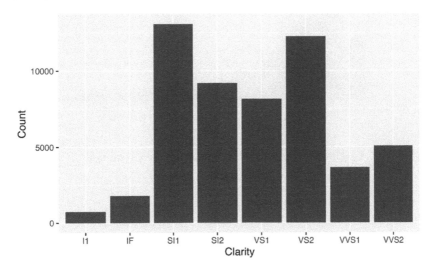

**FIGURE 7.13** Showing the overall proportions for each color

### 7.6.2 Visualizing distributions with geom_histogram(), geom_freqpoly(), and geom_density()

The histogram is an important plot for exploring a continuous variable and counting the number of occurances within certain ranges, or bins. The function geom_histogram() is used to generate a histogram, and the arguments bins and bin_width can be used to specify either the number of bins, or the fixed width of a bin. Below, we specify the total number of bins to be (an arbitrary) value of 15, and the resulting histogram is displayed in Figure 7.14.

```
ggplot(data=diamonds,mapping=aes(x=price))+
  geom_histogram(bins = 15)
```

Another way to present count data is to display it using lines, and this is useful if more that one set of count data needs to be visualized. For this, the function geom_freqpoly() can be used to show the separate counts for each type of clarity, and these are shown in Figure 7.15.

```
ggplot(data=diamonds,mapping=aes(x=table,color=clarity))+
  geom_freqpoly()
```

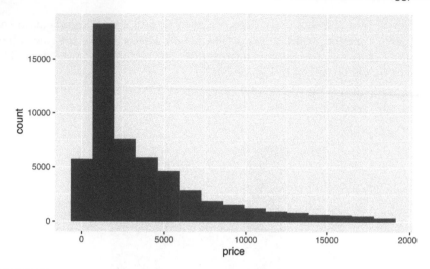

**FIGURE 7.14** Display the price histogram for all diamonds

```
#> `stat_bin()` using `bins = 30`. Pick better value with
#> `binwidth`.
```

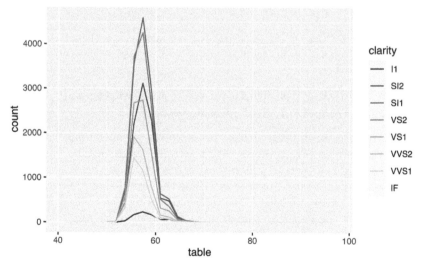

**FIGURE 7.15** Showing the frequency counts using lines

Finally, it is always valuable to display the kernel density estimate (a smoothed version if the histogram) with the function geom_density(), as shown below, and visualized in Figure 7.16.

```
ggplot(data=diamonds,mapping=aes(x=price))+
  geom_density()
```

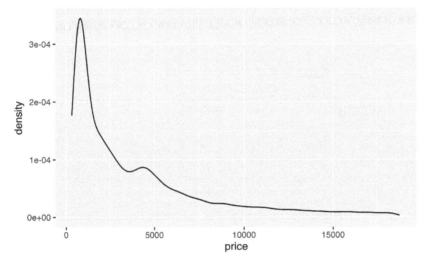

**FIGURE 7.16** Showing the kernel density estimate

### 7.6.3 Exploring the five-number summary with `geom_boxplot()`

The boxplot is a valuable graph that visualizes the distribution of a continuous variable, showing the median, the 25th to 75th percentiles, with lines drawn to the position of the value less than or equal to 1.5 times the interquartile range (IQR). Values outside 1.5 times the IQR are outliers, and shown as points. Because of the wider range of distributions in the mpg dataset, we show the cty continuous variable for each car class. As expected, it shows lower median values of cty for cars with larger engines (e.g., pickups) and higher median values for the smaller cars (e.g., compact). The boxplot is shown in Figure 7.17.

```
ggplot(data=mpg,mapping=aes(y=cty,x=class))+
  geom_boxplot()
```

### 7.6.4 Covariation with `ggpairs()`

The package GGally contains the function ggpairs which visualizes relationships between variables, presents the density plot for each variable, and summarizes the range of correlation coefficients between continuous variables. It provides a useful perspective on the data, and for this example, we use the mpg dataset and the three variables cty, hwy, and displ. We will make use of this plot for exploratory data analysis in Chapter 12, when we explore the associations

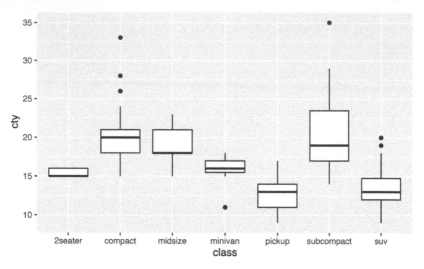

**FIGURE 7.17** Highlighting continuous variables using a boxplot

between variables in a housing dataset. Below we show the call to `ggpairs()`, based on data contained in the tibble `mpg`, and the output is visualized in Figure 7.18.

```
library(GGally)
my_vars <- subset(mpg,select=c(cty,hwy,displ))
ggpairs(my_vars)
```

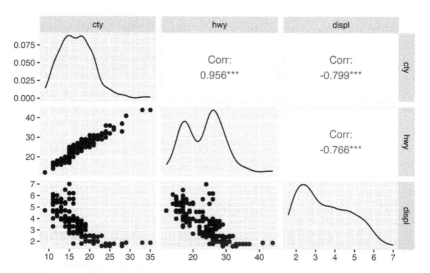

**FIGURE 7.18** Visualizing possible associations between three variables

## 7.7 Themes

The ggplot2 theme system enables you to control non-data elements of your plot, for example, control over elements such as fonts, ticks, legends, and backgrounds. It comprises the following components (Wickham, 2016):

- Theme elements specify the non-data elements; for example, legend.title controls the appearance of the legend title.
- Element functions, which describe the visual properties of an element. For example, element_text() can be used to set the font size, color, and typeface of text elements such as those used in legend.title. Other element functions include element_rect() and element_line().
- The theme() function where you can change the default theme elements by calling element functions, for example, theme(legend.position="top").
- Complete and ready-to-use themes, such as theme_bw() and theme_light(), which are designed to provide a coherent set of values to work with your plots. Themes are also available from within the package ggthemes, which includes theme_economist(), based on the style of *The Economist* newspaper.

To explore an example of how this theme system can be used, we will revert to an earlier plot, and make five arbitrary changes:

- The font type of the legend will be changed to bold face, using the argument legend.title and the element function element_text().
- The legend will be moved to the top of the chart, based on setting the argument legend.position to "top".
- The text on the x-axis will be made larger with a new font, by setting argument axis.text.x using the element function element_text().
- The plot background is altered through the use of the argument panel.background and the element function element_rect().
- The appearance of the grid is changed through the argument panel.grid and the element function element_line().

The modified plot is shown in Figure 7.19.

```
ggplot(data=mpg,aes(x=displ,y=cty,color=class))+
  geom_point()+
  theme(legend.title      = element_text(face="bold"),
        legend.position   = "top",
        axis.text.x       = element_text(size=15,face="italic"),
        panel.background  = element_rect(fill="white"),
        panel.grid        = element_line(color = "blue",
                                         linewidth = 0.3,
                                         linetype = 3))
```

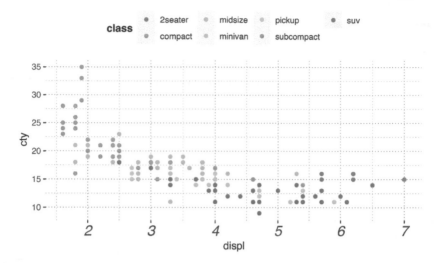

**FIGURE 7.19** Experimenting with theme() to change the plot appearance

Note that there are many other options that can be used within the `theme()`
function, and for full details type `?theme` in the R console. The default theme in
`ggplot2()` can be also invoked using `theme_gray()`, and there are other themes
that can be used, including:

- `theme_bw()`, a classic dark-on-light theme.
- `theme_light()`, a theme with light gray lines and axes in order to provide
  more emphasis on the data.
- `theme_dark()`, referred to in the documentation as the dark cousin of
  theme_light().
- `theme_minimal()`, a minimalistic theme with no background annotations.
- `theme_classic()`, a classic-looking theme, with no gridlines.
- `theme_void()`, a completely empty theme.

To compare these themes, we create a new tibble with just five data points,
and generate an object `plot` that can store the basic scatterplot.

```
d <- tibble(x=seq(1,3,by=0.5),y=2*x)
plot <- ggplot(data=d,mapping=aes(x=x,y=y))+
        geom_point()
```

Figure 7.20 shows six different plots based on which of the six themes is invoked.
The choice of theme is fully up to the analyst, for example, there may be
specific themes that may have to be used, or it may depend on the purpose
of the plot (it could be required for a publication, and therefore certain plot
attributes may have to be adhered to). A benefit of this approach is that
additional themes, for example those contained in the package `ggthemes`, can
also be used.

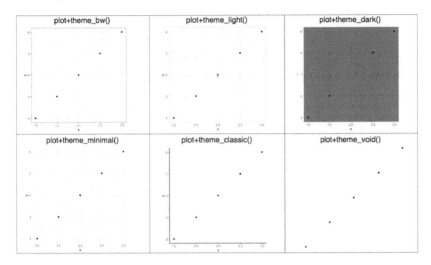

**FIGURE 7.20** Several of the themes available in ggplot2

## 7.8 Adding lines to a plot

There may be situations where adding a layer containing a line is required for your plot. This could be a vertical, horizontal, or a line with a slope. For example, on a histogram, you may want to add vertical lines representing the median and 95% quantiles. There are three functions for doing this, specifically:

- geom_vline(), where the argument xintercept is fixed to position the vertical line.
- geom_hline(), with the argument yintercept used to position the horiztonal line.
- geom_abline(), where both the arguments slope and intercept can be set to generate the output line. For example, this is a convenient function if you want to display your own line of best fit based on outputs from lm().

A sample plot showing how these three functions can be used is shown in Figure 7.21. There is flexibility to also change the line appearance through the argument linetype.

```
ggplot(data=mpg,mapping=aes(x=displ,y=cty))+
  geom_point()+
  geom_vline(xintercept = 4,color="red")+
  geom_hline(yintercept = 22.5,color="blue")+
  geom_abline(slope=10,intercept=-17.5,linetype="dashed")
```

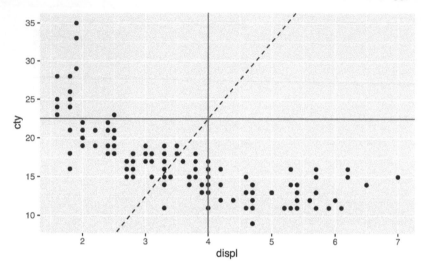

**FIGURE 7.21** Adding lines to a plot

## 7.9    Mini-case: Visualizing the impact of Storm Ophelia

In this mini-case, we take our first look at the dataset `aimsir17`, which contains hourly observations from 25 weather stations in Ireland, all recorded in 2017. *Aimsir* is the Irish word for weather, and was also a convenient choice for the package name as it ends in the letter "r". There are 219,000 records in all (25 stations × 365 days × 24 hourly observations), and here is a look at the first ten observations. Data recorded includes the month, day, hour, time, rainfall, temperature, humidity, mean sea level pressure, wind speed, and wind direction. Note that the `date` variable is a special R data format that is recognized by `ggplot2` for time series plots.

```
library(aimsir17)
```

```
observations
#> # A tibble: 219,000 x 12
#>     station  year  month    day  hour date                   rain
#>     <chr>    <dbl> <dbl>  <int> <int> <dttm>                 <dbl>
#>   1 ATHENRY  2017      1      1     0 2017-01-01 00:00:00        0
#>   2 ATHENRY  2017      1      1     1 2017-01-01 01:00:00        0
#>   3 ATHENRY  2017      1      1     2 2017-01-01 02:00:00        0
#>   4 ATHENRY  2017      1      1     3 2017-01-01 03:00:00      0.1
#>   5 ATHENRY  2017      1      1     4 2017-01-01 04:00:00      0.1
#>   6 ATHENRY  2017      1      1     5 2017-01-01 05:00:00        0
#>   7 ATHENRY  2017      1      1     6 2017-01-01 06:00:00        0
```

```
#>  8 ATHENRY  2017      1     1     7 2017-01-01 07:00:00   0
#>  9 ATHENRY  2017      1     1     8 2017-01-01 08:00:00   0
#> 10 ATHENRY  2017      1     1     9 2017-01-01 09:00:00   0
#> # ... with 218,990 more rows, and 5 more variables: temp <dbl>,
#> #   rhum <dbl>, msl <dbl>, wdsp <dbl>, wddir <dbl>
```

We first need to "drill down" and generate a reduced dataset for the three days in question, and we will also select a number of weather stations from different parts of Ireland. These are *Belmullet* (north west), *Roches Point* (south), and *Dublin Airport* (east). Note that the station names in the tibble are all upper-case. The following code uses the function `subset()` to filter the dataset, and the & operator is used as we have more than one filtering condition. We select columns that contain the date, temperature, mean sea level pressure, and wind speed. In Chapter 8 we will start using the functions from `dplyr` to filter data, and we will no longer need to use `subset()`.

```
storm <- observations |>
         subset(month==10 &
                day %in% 15:17 &
                station %in% c("BELMULLET",
                               "ROCHES POINT",
                               "DUBLIN AIRPORT"),
                select=c(station,date,temp,msl,wdsp))
storm
#> # A tibble: 216 x 5
#>    station    date                 temp   msl  wdsp
#>    <chr>      <dttm>              <dbl> <dbl> <dbl>
#>  1 BELMULLET  2017-10-15 00:00:00  15.5 1004.    30
#>  2 BELMULLET  2017-10-15 01:00:00  15.9 1003.    33
#>  3 BELMULLET  2017-10-15 02:00:00  15.3 1002.    33
#>  4 BELMULLET  2017-10-15 03:00:00  15.1 1002.    35
#>  5 BELMULLET  2017-10-15 04:00:00  15.1 1001.    33
#>  6 BELMULLET  2017-10-15 05:00:00  15.4 1001.    34
#>  7 BELMULLET  2017-10-15 06:00:00  15    1002.    32
#>  8 BELMULLET  2017-10-15 07:00:00  13.7 1003.    28
#>  9 BELMULLET  2017-10-15 08:00:00  11.7 1006.    17
#> 10 BELMULLET  2017-10-15 09:00:00  11.1 1008.    13
#> # ... with 206 more rows
```

We can now use our knowledge of `ggplot2` to explore these weather variables and gain an insight into how the storm evolved over the 3 days in October 2017. This weather data is convenient, as we can use it to explore `ggplot2` in a practical way.

- *Exploration one*, where we plot, over time, the mean sea level pressure for each weather station, and this is shown in Figure 7.22. With stormy weather systems, you would expect to see a significant drop in mean sea level pressure. This is evident from the plot, with the the lowest vaues coinciding with the peak of the storm. Here we also call set the argument `axis.text.x` to show how we can rotate the axis text by 90 degrees.

```
ggplot(data=storm,aes(x=date,y=msl,color=station))+
  geom_point()+
  geom_line() +
  theme(legend.position = "top",
        axis.text.x = element_text(angle = 90))+
  scale_x_datetime(date_breaks = "8 hour",date_labels = "%H:%M %a")+
  labs(title="Storm Ophelia",
       subtitle = "Mean Sea Level Pressure",
       x="Day and Time",
       y="Mean Sea Level Pressure (hPa)",
       color="Weather Station")
```

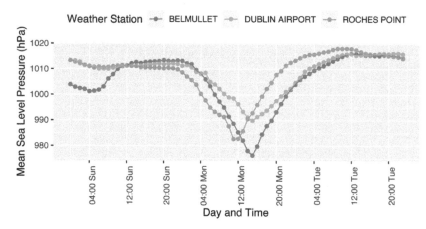

**FIGURE 7.22** Mean sea level pressure during Storm Ophelia

When plotting time series (where you have an R date object), it is worth exploring how to use the function `scale_x_datetime()`, as it provides an excellent way to format dates. Here we use two arguments. The first is `date_breaks` which specifies the x-axis points that will contain date labels (every 8 hours), and the second is `date_labels` which allows you to configure what will be printed (hour, minute, and day of week).

A range of values can be extracted, and configured, and the table below (Wickham, 2016) shows how the special character % can be used in combination with a character. For example, '%H:%M %a" will combine the hour (24-hour clock), a colon, the minute (00–59), a blank space followed by the abbreviated day of the week. It allows you to present a user-friendly and readable x-axis, and the important point is that these labels are fully configurable.

| String | Meaning for `date_labels` |
|--------|---------------------------|
| %S | Second (00–59) |
| %M | Minute (00–59) |
| %l | Hour, 12 hour clock (1–12) |
| %I | Hour, 12 hour clock (01–12) |
| %p | am/pm |
| %H | Hour, 24-hour clock (00–23) |
| %a | Day of week, abbreviated (Mon–Sun) |
| %A | Day of week, full (Monday–Sunday) |
| %e | Day of month (1–31) |
| %d | Day of month (00–31) |
| %m | Month, numeric (01–12) |
| %b | Month, abbreviated (Jan–Dec) |
| %B | Month, full (January–December) |
| %y | Year, without century (00–99) |
| %Y | Year, with century (0000–9999) |

- *Exploration two*, where we plot, over time, the wind speed (knots) for each weather station. This shows an interesting variation in the average hourly wind speed patterns across the three stations during the day, with the maximum mean wind speed recorded in Roches Point, which is just south of Cork City. The data indicates that the southerly points in Ireland were hardest hit by the storm. These wind speed values are visualized in Figure 7.23.

```
ggplot(data=storm,aes(x=date,y=wdsp,color=station))+
  geom_point()+
  geom_line()+
  theme(legend.position = "top",
        axis.text.x = element_text(angle = 90))+
  scale_x_datetime(date_breaks = "8 hour",date_labels = "%H:%M %a")+
  labs(title="Storm Ophelia",
       subtitle = "Wind Speed",
       x="Day and Time",
       y="Wind Speed (Knots)",
       color="Weather Station")
```

- *Exploration three*, which is somewhat speculative but still interesting, as we plot the relationship between the mean sea level pressure and the wind speed, taken over the 3 days in October. We color this by station, and also layer a linear model for each of the three stations. The linear model is visualized using a call to `geom_smooth(method="lm")`, and this is a powerful feature of

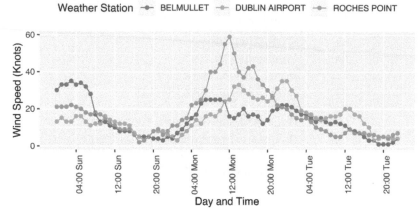

**FIGURE 7.23** Wind speed during Storm Ophelia

ggplot2 that allows rapid model results to be visualized. In exploratory data analysis, this is a good approach as we can get a sense of whether there may be linear relationships between variables. In this case, and based on the small dataset, an increase in atmospheric pressure does seem negatively correlated with the average wind speed, and this is visualized in Figure 7.24.

**Storm Ophelia**
Atmospheric Pressure v Wind Speed with geom_smooth()

Weather Station — BELMULLET — DUBLIN AIRPORT — ROCHES POINT

**FIGURE 7.24** Wind speed during Storm Ophelia

```
ggplot(data=storm,aes(x=msl,y=wdsp,color=station))+
  geom_point()+
  geom_smooth(method="lm")+
  theme(legend.position = "top")+
  labs(title="Storm Ophelia",
      subtitle="Atmospheric Pressure v Wind Speed with geom_smooth()",
      x="Mean Sea Level Pressure (hPa)",
      y="Wind Speed (Knots)",
      color="Weather Station")
```

Further analysis could be performed on the `aimsir17` dataset, for example, by locating other storm systems from 2017, finding related media content, and also observing how variables such as wind speed and mean sea level pressure varied over time.

## 7.10   Summary of R functions from Chapter 7

Many important functions from the package `ggplot2` are now summarized. Functions are assumed to be part of `ggplot2`, unless otherwise indicated.

| Function | Description |
| --- | --- |
| aes() | Maps variables to visual properties of geoms. |
| element_line() | Specify properties of a line, used with theme(). |
| element_rect() | Specify properties of borders and backgrounds. |
| element_text() | Specify properties of text. |
| facet_wrap() | Creates panels based on an input variable |
| facet_grid() | Creates a matrix of panels for two variables. |
| ggplot() | Initializes a ggplot object. |
| geom_ point() | Used to create scatterplots. |
| geom_abline() | Draws a line with a given slope and intercept. |
| geom_bar() | Constructs a bar chart based on count data. |
| geom_boxplot() | Displays summary of distribution. |
| geom_density() | Computes and draws the kernel density estimate. |
| geom_freqpoly() | Visualize counts of one or more variables. |
| geom_histogram() | Visualize the distribution of a variable. |
| geom_hline() | Draws a horizonal line. |
| geom_vline() | Draws a vertical line. |
| geom_smooth() | Visualization aid to assist in showing patterns. |
| ggpairs() | Creates a matrix of plots (library GGally). |
| labs() | Used to modify axis, legend and plot labels. |

| Function | Description |
|---|---|
| scale_x_continuous() | Configure the x-axis scales. |
| scale_y_continuous() | Configure the y-axis scales. |
| scale_x_datetime() | Used to configure date objects on the x-axis. |
| scale_y_datetime() | Configure date objects on the y-axis. |
| theme() | Customize the non-data components of your plots. |
| theme_grey() | Signature theme for ggplot2. |
| theme_bw() | Dark-on-light theme. |
| theme_light() | Theme with light gray lines and axes. |
| theme_dark() | Similar to theme_light(), darker background. |
| theme_minimal() | Minimalistic theme. |
| theme_classic() | Theme with no background annotations. |
| theme_void() | Completely empty theme. |

## 7.11  Exercises

1.  Generate the following plot from the mpg tibble in ggplot2. The
    x-variable is displ and the y-variable cty. Make use of the lab()
    and theme() functions.

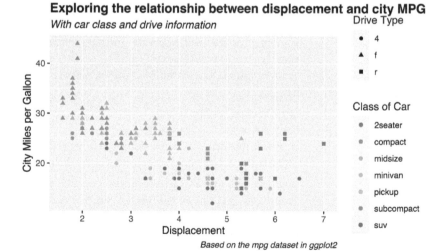

2.  Use ggpairs() to explore a subset of the Irish weather dataset
    (aimsir17), and the possible relationships between the variables
    rain, wdsp, temp and msl. Use the station "ROCHES POINT" and

all the observations from the month of October (month 10), which can be retrieved using the `subset()` function.

**Exploring relationships between Irish weather variables**

*Rainfall, wind speed, temperature and mean sea level pressure*

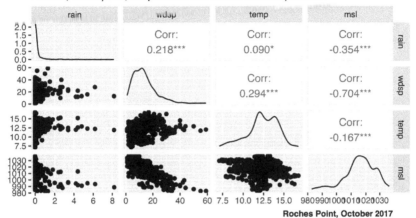

**Roches Point, October 2017**

3. Using the function `subset()`, extract observations for three weather stations *Dublin Airport*, *Malin Head*, and *Belmullet*, and generate the following comparative temperature box plot summaries over the twelve months of the year. In the plot, convert the month variable to a factor value, using the function `factor()`. Note that the station names are stored as uppercase.

**Summaries of monthly temperature in 2017**

*For stations Dublin Airport (E), Malin Head (N) and Sherkin Island (S)*

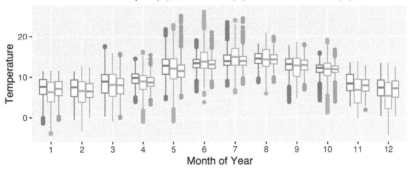

*Based on the observations dataset in aimsir17*

# 8

## Data Transformation with `dplyr`

> Writing code with `dplyr` involves using the "grammar of data" to perform data munging. Each step is done by a single function that represents a verb.
>
> — Jared P. Lander (Lander, 2017)

## 8.1 Introduction

Visualization is an important way to explore and gain insights from data. However, a tibble may not contain the data that is required, and further transformation of the tibble may be needed. In earlier chapters, functions such as `subset()` and `transform()` were used to (1) select rows according to a conditional expression, (2) select columns that are of interest, and (3) add new columns that are functions of other columns. The package `dplyr` provides an alternative set of functions to support data transformations, and also additional functions that can be used to summarize data. The underlying structure of `dplyr` is often termed "tibble-in tibble out" (Wickham and Grolemund, 2016), where each function accepts a tibble and a number of arguments, and then returns a tibble. This provides an elegant architecture that is ideal for the use of tools such as the R and tidyverse pipe operators. Note that in our presentation of code examples from the tidyverse, we have opted to use the full reference for functions so as to be clear which package is being used (this will be optional for you as you code your own solutions).

Upon completing the chapter, you should understand the following tidyverse functions:

- The tidyverse pipe operator %>% (from the package magrittr), which allows you to chain together a sequence of functions that transform data.
- The dplyr functions filter(), arrange(), select(), mutate(), group_by(), and summarize(), all of which allow you to manipulate and transform tibbles in a structured way.
- Additional dplyr functions pull() and case_when().
- A summary of R functions that allow you to use the tools of dplyr.
- How to solve all five test exercises.

## Chapter structure

- Section 8.2 introduces the tidyverse pipe %>%, which is similar to the R native pipe |>.
- Section 8.3 describes the function filter(), which allows you to subset the rows of a tibble.
- Section 8.4 presents the function arrange() which is used to reorder a tibble (ascending or descending) based on one or more variables.
- Section 8.5 summarizes the function select(), which allows you to select a subset of columns from a tibble.
- Section 8.6 describes the function mutate(), which is used to add new columns to a tibble.
- Section 8.7 presents the function summarize() which allows you to obtain summaries of variable groupings, for example, summations, averages, and standard deviations.
- Section 8.8 documents a number of other functions that integrate well with dplyr.
- Section 8.9 presents a study of how we can use dplyr to compute summary rainfall information from the package aimsir17.
- Section 8.10 provides a summary of the functions introduced in the chapter.
- Section 8.11 provides a number of short coding exercises to test your understanding of dplyr.

## 8.2   The tidyverse pipe from magrittr

Earlier in Chapter 4, we introduced the native pipe operator in R |>, which allowed for the chaining together of a number of functions, where the output from one function can be "piped" into the input of another. A similar pipe operator is also available in the tidyverse and is represented by the symbol %>%. This operator is contained in the package magrittr and is also loaded by default when using the package dplyr. While there are several differences between these two pipes, the way in which we use them in this textbook is equivalent. To explore the benefits of using the pipe operator, consider the following short example. First, we load the package `magrittr`.

```
library(magrittr)
```

Next, we implement the following workflow that could have applications in simulating customer arrivals at a restaurant over a 12-hour period. Assume the arrivals follow a Poisson distribution with a mean of 50 per hour. We want to store the top six customer values drawn. This could be done without using a pipe, with the following code.

```
set.seed(100)
arr    <- rpois(12,50)
arr
#>  [1] 46 38 56 50 52 45 53 47 50 50 48 55
# The function sort takes a parameter to sort in descending order
s_arr <- sort(arr,decreasing = TRUE)
top_6 <- head(s_arr)
top_6
#> [1] 56 55 53 52 50 50
```

With the `%>%` operator, this sequential workflow can be simplifed into fewer steps, and note the benefit is that the number of variables needed is reduced. In order to keep the code as clear as possible, each function is shown on a separate line.

```
set.seed(100)
top_6   <- rpois(12,50) %>%
            sort(decreasing = TRUE) %>%
            head()
top_6
#> [1] 56 55 53 52 50 50
```

While the pipe operator(s) are valuable and can lead to cleaner and easier to understand code, it is important to note that it is best to use a short number of steps (not more that ten is a useful guide) where the output from one is the input to the next. Also, the process is not designed for multiple inputs or outputs. However, despite these constraints, it is a wonderful operator, and we will make extensive use of this as we study the `tidyverse`.

---

## 8.3 Filtering rows with `filter()`

The function `filter()` is used to subset a tibble, in which all rows satisfying a condition are retained. It is an important function for data manipulation, as often you may need to "zoom in" on a subset of data from a larger dataset, and perform operations on this reduced set of rows. This function conforms to the *architecture* of dplyr, in that it accepts a tibble, and arguments, and always returns a tibble. The arguments include:

- A data frame, or data frame extensions such as a tibble.
- A list of expressions that return a logical value and are defined in terms of variables that are present in the input data frame. More than one expression can be provided, and these will act as separate filtering conditions.

To explore filter, let's consider the mpg data frame, and here we can filter all vehicles that belong to the class "2seater". We store the result in a new tibble called mpg1, which is now displayed.

```
mpg1 <- dplyr::filter(mpg,class=="2seater")
mpg1
#> # A tibble: 5 x 11
#>   manufac~1 model displ  year   cyl trans drv      cty    hwy fl
#>   <chr>     <chr> <dbl> <int> <int> <chr> <chr> <int> <int> <chr>
#> 1 chevrolet corv~   5.7  1999     8 manu~ r        16     26 p
#> 2 chevrolet corv~   5.7  1999     8 auto~ r        15     23 p
#> 3 chevrolet corv~   6.2  2008     8 manu~ r        16     26 p
#> 4 chevrolet corv~   6.2  2008     8 auto~ r        15     25 p
#> 5 chevrolet corv~   7    2008     8 manu~ r        15     24 p
#> # ... with 1 more variable: class <chr>, and abbreviated
#> #   variable name 1: manufacturer
```

Notice that the result stored in mpg1 has retained all of the columns, and those rows (5 in all) that match the condition. The condition explicitly references the column value that must be true for all matching observations. We can add more conditions for the subsetting task, for example, to include those values of the "2seater" class, where the value for highway miles per gallon (hwy) is greater than or equal to 25.

```
mpg2 <- dplyr::filter(mpg,class=="2seater",hwy >= 25)
mpg2
#> # A tibble: 3 x 11
#>   manufac~1 model displ  year   cyl trans drv      cty    hwy fl
#>   <chr>     <chr> <dbl> <int> <int> <chr> <chr> <int> <int> <chr>
#> 1 chevrolet corv~   5.7  1999     8 manu~ r        16     26 p
#> 2 chevrolet corv~   6.2  2008     8 manu~ r        16     26 p
#> 3 chevrolet corv~   6.2  2008     8 auto~ r        15     25 p
#> # ... with 1 more variable: class <chr>, and abbreviated
#> #   variable name 1: manufacturer
```

As can be seen with the new tibble mpg2, this reduces the result to just three observations. But say if you wanted to view all cars that were manufactured by either "lincoln" or "mercury", a slightly different approach is needed. This is an "or" operation, where we want to return observations that match one of these two values. To achieve this, we can compose a conditional expression with two elements, and connected by R's | operator. The results are shown in the tibble mpg3.

```
mpg3 <- dplyr::filter(mpg,
                manufacturer == "lincoln" |
                manufacturer == "mercury")
mpg3
#> # A tibble: 7 x 11
#>   manufac~1 model displ  year   cyl trans drv     cty   hwy fl
#>   <chr>     <chr> <dbl> <int> <int> <chr> <chr> <int> <int> <chr>
#> 1 lincoln   navi~   5.4  1999     8 auto~ r        11    17 r
#> 2 lincoln   navi~   5.4  1999     8 auto~ r        11    16 p
#> 3 lincoln   navi~   5.4  2008     8 auto~ r        12    18 r
#> 4 mercury   moun~   4    1999     6 auto~ 4        14    17 r
#> 5 mercury   moun~   4    2008     6 auto~ 4        13    19 r
#> 6 mercury   moun~   4.6  2008     8 auto~ 4        13    19 r
#> 7 mercury   moun~   5    1999     8 auto~ 4        13    17 r
#> # ... with 1 more variable: class <chr>, and abbreviated
#> #   variable name 1: manufacturer
```

A preferable alternative to using the | operator is to use the %in% operator which can be used to identify whether an element is part of a vector. In this case, the two manufacturers ("lincoln" and "mercury") are added to an atomic vector, and that is used as part of the %in% operator to locate the rows.

```
mpg4 <- dplyr::filter(mpg,
                manufacturer %in% c("lincoln","mercury"))
mpg4
#> # A tibble: 7 x 11
#>   manufac~1 model displ  year   cyl trans drv     cty   hwy fl
#>   <chr>     <chr> <dbl> <int> <int> <chr> <chr> <int> <int> <chr>
#> 1 lincoln   navi~   5.4  1999     8 auto~ r        11    17 r
#> 2 lincoln   navi~   5.4  1999     8 auto~ r        11    16 p
#> 3 lincoln   navi~   5.4  2008     8 auto~ r        12    18 r
#> 4 mercury   moun~   4    1999     6 auto~ 4        14    17 r
#> 5 mercury   moun~   4    2008     6 auto~ 4        13    19 r
#> 6 mercury   moun~   4.6  2008     8 auto~ 4        13    19 r
#> 7 mercury   moun~   5    1999     8 auto~ 4        13    17 r
#> # ... with 1 more variable: class <chr>, and abbreviated
#> #   variable name 1: manufacturer
```

Note that in practice, the %in% operator returns a logical vector which is then used as an argument to filter. This process can be made more explicit; for example, consider the following code whereby we can extract the first record from the tibble mpg using a logical vector (which could have also been created using %in%).

```
lv <- c(T,rep(F,nrow(mpg)-1))
dplyr::filter(mpg,lv)
#> # A tibble: 1 x 11
#>   manufac~1 model displ  year   cyl trans drv     cty   hwy fl
#>   <chr>     <chr> <dbl> <int> <int> <chr> <chr> <int> <int> <chr>
#> 1 audi      a4      1.8  1999     4 auto~ f        18    29 p
#> # ... with 1 more variable: class <chr>, and abbreviated
#> #   variable name 1: manufacturer
```

A function that can be used to subset a data frame or tibble is slice(), which allows you to index tibble rows by their (integer) locations.

```
# Show the first 3 rows
dplyr::slice(mpg,1:3)
#> # A tibble: 3 x 11
#>   manufac~1 model displ  year   cyl trans drv     cty   hwy fl
#>   <chr>     <chr> <dbl> <int> <int> <chr> <chr> <int> <int> <chr>
#> 1 audi      a4      1.8  1999     4 auto~ f        18    29 p
#> 2 audi      a4      1.8  1999     4 manu~ f        21    29 p
#> 3 audi      a4      2    2008     4 manu~ f        20    31 p
#> # ... with 1 more variable: class <chr>, and abbreviated
#> #   variable name 1: manufacturer
# Sample 3 rows
set.seed(100)
dplyr::slice(mpg,sample(1:nrow(mpg),size = 3))
#> # A tibble: 3 x 11
#>   manufac~1 model displ  year   cyl trans drv     cty   hwy fl
#>   <chr>     <chr> <dbl> <int> <int> <chr> <chr> <int> <int> <chr>
#> 1 toyota    toyo~   2.7  1999     4 auto~ 4        16    20 r
#> 2 honda     civic   1.6  1999     4 manu~ f        25    32 r
#> 3 hyundai   sona~   2.4  2008     4 manu~ f        21    31 r
#> # ... with 1 more variable: class <chr>, and abbreviated
#> #   variable name 1: manufacturer
```

## 8.4   Sorting rows with arrange()

The arrange() function orders the rows of a tibble by the values of selected columns, and these can be in ascending order (the default), or descending order, by using the function desc(). For example, we can order the tibble mpg by the variable cty, and we show the first three rows.

```
dplyr::arrange(mpg,cty) %>% slice(1:3)
#> # A tibble: 3 x 11
#>    manufac~1 model displ year   cyl trans drv     cty   hwy fl
#>    <chr>     <chr> <dbl> <int> <int> <chr> <chr> <int> <int> <chr>
#> 1 dodge     dako~   4.7  2008     8 auto~ 4         9    12 e
#> 2 dodge     dura~   4.7  2008     8 auto~ 4         9    12 e
#> 3 dodge     ram ~   4.7  2008     8 auto~ 4         9    12 e
#> # ... with 1 more variable: class <chr>, and abbreviated
#> #    variable name 1: manufacturer
```

We can then re-order `mpg` by `cty`, from highest to lowest, by wrapping the column name with the function `desc()`.

```
dplyr::arrange(mpg,desc(cty)) %>% slice(1:3)
#> # A tibble: 3 x 11
#>    manufac~1 model displ year   cyl trans drv     cty   hwy fl
#>    <chr>     <chr> <dbl> <int> <int> <chr> <chr> <int> <int> <chr>
#> 1 volkswag~ new ~   1.9  1999     4 manu~ f        35    44 d
#> 2 volkswag~ jetta   1.9  1999     4 manu~ f        33    44 d
#> 3 volkswag~ new ~   1.9  1999     4 auto~ f        29    41 d
#> # ... with 1 more variable: class <chr>, and abbreviated
#> #    variable name 1: manufacturer
```

More than one column can be provided for re-ordering data. For example, we can re-order `mpg` by `class`, and then by `hwy`. Notice that it will take the first class (five observations), and order those rows, before moving on to the next class.

```
dplyr::arrange(mpg,class,desc(cty)) %>% slice(1:7)
#> # A tibble: 7 x 11
#>    manufac~1 model displ year   cyl trans drv     cty   hwy fl
#>    <chr>     <chr> <dbl> <int> <int> <chr> <chr> <int> <int> <chr>
#> 1 chevrolet corv~   5.7  1999     8 manu~ r        16    26 p
#> 2 chevrolet corv~   6.2  2008     8 manu~ r        16    26 p
#> 3 chevrolet corv~   5.7  1999     8 auto~ r        15    23 p
#> 4 chevrolet corv~   6.2  2008     8 auto~ r        15    25 p
#> 5 chevrolet corv~   7    2008     8 manu~ r        15    24 p
#> 6 volkswag~ jetta   1.9  1999     4 manu~ f        33    44 d
#> 7 toyota    coro~   1.8  2008     4 manu~ f        28    37 r
#> # ... with 1 more variable: class <chr>, and abbreviated
#> #    variable name 1: manufacturer
```

## 8.5   Choosing columns with `select()`

The function `select()` allows you to subset columns from the input tibble
using the column names. In situations where there are many columns in a data
frame, this is an important function. Note that `select()` will always return all
of the observations for the selected columns. The simplest use of `select()` is
to write the column names that are required.

```
dplyr::select(mpg,manufacturer,model,year,displ,cty)
#> # A tibble: 234 x 5
#>     manufacturer model       year displ   cty
#>     <chr>        <chr>      <int> <dbl> <int>
#>  1 audi         a4          1999   1.8    18
#>  2 audi         a4          1999   1.8    21
#>  3 audi         a4          2008   2      20
#>  4 audi         a4          2008   2      21
#>  5 audi         a4          1999   2.8    16
#>  6 audi         a4          1999   2.8    18
#>  7 audi         a4          2008   3.1    18
#>  8 audi         a4 quattro  1999   1.8    18
#>  9 audi         a4 quattro  1999   1.8    16
#> 10 audi         a4 quattro  2008   2      20
#> # ... with 224 more rows
```

The function also provides a number of operators that make it easier to select
variables, for example:

- `:` for selecting a range of consecutive variables

```
mpg %>% dplyr::select(manufacturer:year,cty)
#> # A tibble: 234 x 5
#>     manufacturer model      displ  year   cty
#>     <chr>        <chr>      <dbl> <int> <int>
#>  1 audi         a4          1.8   1999    18
#>  2 audi         a4          1.8   1999    21
#>  3 audi         a4          2     2008    20
#>  4 audi         a4          2     2008    21
#>  5 audi         a4          2.8   1999    16
#>  6 audi         a4          2.8   1999    18
#>  7 audi         a4          3.1   2008    18
#>  8 audi         a4 quattro  1.8   1999    18
#>  9 audi         a4 quattro  1.8   1999    16
#> 10 audi         a4 quattro  2     2008    20
#> # ... with 224 more rows
```

- ! for taking the complement (i.e., negating) a collection of variables

```
mpg %>% dplyr::select(!(manufacturer:year))
#> # A tibble: 234 x 7
#>       cyl trans        drv     cty   hwy fl      class
#>     <int> <chr>        <chr> <int> <int> <chr>   <chr>
#> 1      4 auto(l5)     f       18    29 p       compact
#> 2      4 manual(m5)   f       21    29 p       compact
#> 3      4 manual(m6)   f       20    31 p       compact
#> 4      4 auto(av)     f       21    30 p       compact
#> 5      6 auto(l5)     f       16    26 p       compact
#> 6      6 manual(m5)   f       18    26 p       compact
#> 7      6 auto(av)     f       18    27 p       compact
#> 8      4 manual(m5)   4       18    26 p       compact
#> 9      4 auto(l5)     4       16    25 p       compact
#> 10     4 manual(m6)   4       20    28 p       compact
#> # ... with 224 more rows
```

- c() for combining selections

```
mpg %>% dplyr::select(c(manufacturer:year,cty))
#> # A tibble: 234 x 5
#>    manufacturer model      displ  year   cty
#>    <chr>        <chr>      <dbl> <int> <int>
#> 1  audi         a4           1.8  1999    18
#> 2  audi         a4           1.8  1999    21
#> 3  audi         a4           2    2008    20
#> 4  audi         a4           2    2008    21
#> 5  audi         a4           2.8  1999    16
#> 6  audi         a4           2.8  1999    18
#> 7  audi         a4           3.1  2008    18
#> 8  audi         a4 quattro   1.8  1999    18
#> 9  audi         a4 quattro   1.8  1999    16
#> 10 audi         a4 quattro   2    2008    20
#> # ... with 224 more rows
```

In addition to these operators, there are a number of functions known as *selection helpers*, and these can be a valuable way to select columns, for example:

- `starts_with()`, which takes a string and returns any column that starts with this value

```
mpg %>% dplyr::select(starts_with("m"))
#> # A tibble: 234 x 2
#>    manufacturer model
#>    <chr>        <chr>
#> 1  audi         a4
```

```
#>   2 audi          a4
#>   3 audi          a4
#>   4 audi          a4
#>   5 audi          a4
#>   6 audi          a4
#>   7 audi          a4
#>   8 audi          a4 quattro
#>   9 audi          a4 quattro
#>  10 audi          a4 quattro
#> # ... with 224 more rows
```

- ends_with(), which takes a string and returns any column that ends with this value.

```
mpg %>% dplyr::select(ends_with("l"))
#> # A tibble: 234 x 4
#>    model         displ   cyl fl
#>    <chr>         <dbl> <int> <chr>
#>  1 a4              1.8     4 p
#>  2 a4              1.8     4 p
#>  3 a4              2       4 p
#>  4 a4              2       4 p
#>  5 a4              2.8     6 p
#>  6 a4              2.8     6 p
#>  7 a4              3.1     6 p
#>  8 a4 quattro      1.8     4 p
#>  9 a4 quattro      1.8     4 p
#> 10 a4 quattro      2       4 p
#> # ... with 224 more rows
```

- contains(), which takes a string and returns any column that contains the value.

```
mpg %>% dplyr::select(contains("an"))
#> # A tibble: 234 x 2
#>    manufacturer trans
#>    <chr>        <chr>
#>  1 audi         auto(l5)
#>  2 audi         manual(m5)
#>  3 audi         manual(m6)
#>  4 audi         auto(av)
#>  5 audi         auto(l5)
#>  6 audi         manual(m5)
#>  7 audi         auto(av)
#>  8 audi         manual(m5)
#>  9 audi         auto(l5)
```

```
#> 10 audi          manual(m6)
#> # ... with 224 more rows
```

- `num_range()`, which matches a range that is made up of a prefix (string) followed by numbers. Note that the tibble `billboard` is contained in the library `tidyr`.

```
billboard %>%
  dplyr::select(artist,track,num_range("wk", 1:3)) %>%
  slice(1:5)
#> # A tibble: 5 x 5
#>   artist      track                   wk1   wk2   wk3
#>   <chr>       <chr>                 <dbl> <dbl> <dbl>
#> 1 2 Pac       Baby Don't Cry (Keep...   87    82    72
#> 2 2Ge+her     The Hardest Part Of ...   91    87    92
#> 3 3 Doors Down Kryptonite              81    70    68
#> 4 3 Doors Down Loser                    76    76    72
#> 5 504 Boyz    Wobble Wobble            57    34    25
```

- `matches()` matches a *regular expression* which accepts a string pattern that is used to find possible matches. For example, the character "." will match any character. Therefore, the pattern "wk.3" will find a column that starts with "wk" followed by any character, and then concludes with the character "3".

```
billboard %>%
  dplyr::select(matches("wk.3")) %>%
  slice(1:3)
#> # A tibble: 3 x 7
#>    wk13  wk23  wk33  wk43  wk53  wk63 wk73
#>   <dbl> <dbl> <dbl> <dbl> <dbl> <dbl> <lgl>
#> 1    NA    NA    NA    NA    NA    NA NA
#> 2    NA    NA    NA    NA    NA    NA NA
#> 3    47     6     3    14    49    NA NA
```

- `everything()` matches all variables and is convenient if you need to move some variables to the first few columns, and then add the rest.

```
mpg %>%
  dplyr::select(manufacturer:year,cty,hwy,everything())
#> # A tibble: 234 x 11
#>   manufa~1 model displ  year   cty   hwy   cyl trans drv   fl
#>   <chr>    <chr> <dbl> <int> <int> <int> <int> <chr> <chr> <chr>
#> 1 audi     a4      1.8  1999    18    29     4 auto~ f     p
#> 2 audi     a4      1.8  1999    21    29     4 manu~ f     p
#> 3 audi     a4      2    2008    20    31     4 manu~ f     p
#> 4 audi     a4      2    2008    21    30     4 auto~ f     p
```

```
#>  5 audi       a4      2.8  1999    16    26    6 auto~ f    p
#>  6 audi       a4      2.8  1999    18    26    6 manu~ f    p
#>  7 audi       a4      3.1  2008    18    27    6 auto~ f    p
#>  8 audi       a4 q~   1.8  1999    18    26    4 manu~ 4    p
#>  9 audi       a4 q~   1.8  1999    16    25    4 auto~ 4    p
#> 10 audi       a4 q~   2    2008    20    28    4 manu~ 4    p
#> # ... with 224 more rows, 1 more variable: class <chr>, and
#> #   abbreviated variable name 1: manufacturer
```

## 8.6   Adding columns with `mutate()`

The function `mutate()` adds new variables to a tibble, while keeping the original ones. It also allows for variables to be overwritten (for example, changing the values or the type), and variables can be deleted by setting their values to NULL. The two main arguments used in `mutate()` are:

- The tibble that is to be transformed.
- A set of name-value pairs, where the name provides the column name, and the values can be (1) a vector of length 1 that is subsequently recycled to the tibble length, (2) a vector of the same length as the tibble, and (3) the value NULL which will remove the column.

To explore `mutate()`, we take a reduced version of `mpg`, with six random observations and a smaller set of columns. Note that we use the dplyr function `sample_n()` to generate a sample from `mpg`, from two classes, "compact" and "midsize".

```
set.seed(100)
mpg_m <- mpg %>%
         dplyr::filter(class %in% c("compact","midsize")) %>%
         dplyr::select(manufacturer:year,cty,class) %>%
         dplyr::sample_n(6)
mpg_m
#> # A tibble: 6 x 6
#>    manufacturer model  displ  year   cty class
#>    <chr>        <chr>  <dbl> <int> <int> <chr>
#> 1 volkswagen   jetta  2      1999    21 compact
#> 2 volkswagen   jetta  2.5    2008    21 compact
#> 3 chevrolet    malibu 3.6    2008    17 midsize
#> 4 volkswagen   gti    2      2008    21 compact
#> 5 audi         a4     2      2008    21 compact
#> 6 toyota       camry  3.5    2008    19 midsize
```

Now, we can explore the three types of changes that can be made to `mpg` with `mutate()`.

- A vector of size 1 is added, and this is recycled to the correct length.

```
mpg_m %>% dplyr::mutate(Test="A test")
#> # A tibble: 6 x 7
#>   manufacturer model  displ year   cty class    Test
#>   <chr>        <chr>  <dbl> <int> <int> <chr>    <chr>
#> 1 volkswagen   jetta  2     1999     21 compact  A test
#> 2 volkswagen   jetta  2.5   2008     21 compact  A test
#> 3 chevrolet    malibu 3.6   2008     17 midsize  A test
#> 4 volkswagen   gti    2     2008     21 compact  A test
#> 5 audi         a4     2     2008     21 compact  A test
#> 6 toyota       camry  3.5   2008     19 midsize  A test
```

- A vector of the same length is added, and this vector is based on an existing column. For example, we create a new column `cty_kmh` which converts all `cty` values to their counterpart in kilometers (using the constant 1.6). This example shows an important feature of `mutate()` in that existing variables can be used, and also new variables defined within the function call can also be used in the process (for example, the additional variable `cty_2`).

```
mpg_m %>% dplyr::mutate(cty_kmh=cty*1.6,
                        cty_2=cty_kmh/1.6)
#> # A tibble: 6 x 8
#>   manufacturer model  displ year   cty class   cty_kmh cty_2
#>   <chr>        <chr>  <dbl> <int> <int> <chr>     <dbl> <dbl>
#> 1 volkswagen   jetta  2     1999     21 compact    33.6    21
#> 2 volkswagen   jetta  2.5   2008     21 compact    33.6    21
#> 3 chevrolet    malibu 3.6   2008     17 midsize    27.2    17
#> 4 volkswagen   gti    2     2008     21 compact    33.6    21
#> 5 audi         a4     2     2008     21 compact    33.6    21
#> 6 toyota       camry  3.5   2008     19 midsize    30.4    19
```

- Finally, we can remove a variable from a tibble by setting its value to `NULL`

```
mpg_m %>% dplyr::mutate(class=NULL)
#> # A tibble: 6 x 5
#>   manufacturer model  displ year   cty
#>   <chr>        <chr>  <dbl> <int> <int>
#> 1 volkswagen   jetta  2     1999     21
#> 2 volkswagen   jetta  2.5   2008     21
#> 3 chevrolet    malibu 3.6   2008     17
#> 4 volkswagen   gti    2     2008     21
#> 5 audi         a4     2     2008     21
#> 6 toyota       camry  3.5   2008     19
```

Another function that can be used with `mutate()` is `group_by()` which takes a tibble and converts it to a *grouped tibble*, based on the input variable(s). Computations can then be performed on the grouped data. Here, we show how the earlier `mpg_m` variable is grouped by `class`. Note that the data in the tibble does not change, however, information on the group is added to the tibble, and appears when the tibble is displayed.

```
# Group the tibble by class
mpg_mg <- mpg_m %>% dplyr::group_by(class)
mpg_mg
#> # A tibble: 6 x 6
#> # Groups:   class [2]
#>   manufacturer model  displ  year   cty class
#>   <chr>        <chr>  <dbl> <int> <int> <chr>
#> 1 volkswagen   jetta    2    1999    21 compact
#> 2 volkswagen   jetta    2.5  2008    21 compact
#> 3 chevrolet    malibu   3.6  2008    17 midsize
#> 4 volkswagen   gti      2    2008    21 compact
#> 5 audi         a4       2    2008    21 compact
#> 6 toyota       camry    3.5  2008    19 midsize
```

What actually happens is interesting in that a new attribute (groups) is added to the tibble.

```
attr(mpg_mg,"groups")
#> # A tibble: 2 x 2
#>   class        .rows
#>   <chr>  <list<int>>
#> 1 compact        [4]
#> 2 midsize        [2]
```

This information can also be shown using the function `group_keys()`.

```
dplyr::group_keys(mpg_mg)
#> # A tibble: 2 x 1
#>   class
#>   <chr>
#> 1 compact
#> 2 midsize
```

In this chapter, we will mostly use `group_by()` with the `summarize()` function, but it can also be used with `mutate()`. Consider the example where we want to calculate the maximum `cty` value for each class of car in the variable `mpg_m`. By grouping the tibble (by class), the maximum of each *group* will be calculated.

```
mpg_mg %>% dplyr::mutate(MaxCtyByClass=max(cty))
#> # A tibble: 6 x 7
#> # Groups:   class [2]
```

```
#>    manufacturer model   displ  year   cty class   MaxCtyByClass
#>    <chr>        <chr>   <dbl> <int> <int> <chr>            <int>
#> 1  volkswagen   jetta       2  1999    21 compact             21
#> 2  volkswagen   jetta     2.5  2008    21 compact             21
#> 3  chevrolet    malibu    3.6  2008    17 midsize             19
#> 4  volkswagen   gti         2  2008    21 compact             21
#> 5  audi         a4          2  2008    21 compact             21
#> 6  toyota       camry     3.5  2008    19 midsize             19
```

Groupings can then be removed from a tibble with a call to the function upgroup().

```
mpg_mg %>% dplyr::ungroup()
#> # A tibble: 6 x 6
#>    manufacturer model   displ  year   cty class
#>    <chr>        <chr>   <dbl> <int> <int> <chr>
#> 1  volkswagen   jetta       2  1999    21 compact
#> 2  volkswagen   jetta     2.5  2008    21 compact
#> 3  chevrolet    malibu    3.6  2008    17 midsize
#> 4  volkswagen   gti         2  2008    21 compact
#> 5  audi         a4          2  2008    21 compact
#> 6  toyota       camry     3.5  2008    19 midsize
```

## 8.7 Summarizing observations with `summarize()`

The function `summarize()` creates a new tibble. It will have one (or more) rows for each combination of grouping variable. If there are no groupings, then a single row summary of all observations will be displayed. The arguments passed to `summarize()` include:

- The tibble/data frame, which will usually have group attributes defined.
- Name-value pairs of summary functions, where the name will be that of the column in the result. Summary functions include (Wickham and Grolemund, 2016):

| Type | Examples |
|------|----------|
| Measures of location | mean(), median() |
| Measures of spread | sd(), IQR() |
| Measures of rank | min(), max(), quantile() |
| Measures of position | first(), nth(), last() |
| Counts | n(), n_distinct() |
| Proportions | e.g., sum(x>0)/n() |

We can start with a look at summarizing values in a tibble with no groupings. This information really does not add a huge amount, as this type of summary is already available via the summary() function, although the summarize() function does store the result in a tibble.

```
mpg %>%
    dplyr::summarize(CtyAvr=mean(cty),
                     CtySD=sd(cty),
                     HwyAvr=mean(hwy),
                     HwySD=sd(hwy))
#> # A tibble: 1 x 4
#>    CtyAvr CtySD HwyAvr HwySD
#>     <dbl> <dbl>  <dbl> <dbl>
#> 1    16.9  4.26   23.4  5.95
```

To add value to this process, we can group the tibble by whatever variable (or combination of variables) we are interested in. For this example, we group by class, and this will provide seven rows of output, one for each class. Note that the function n() returns the number of observations found for each grouping.

```
mpg %>%
    dplyr::group_by(class) %>%
    dplyr::summarize(CtyAvr=mean(cty),
                     CtySD=sd(cty),
                     HwyAvr=mean(hwy),
                     HwySD=sd(hwy),
                     N=dplyr::n()) %>%
    ungroup()
#> # A tibble: 7 x 6
#>    class      CtyAvr CtySD HwyAvr HwySD     N
#>    <chr>       <dbl> <dbl>  <dbl> <dbl> <int>
#> 1 2seater      15.4 0.548   24.8  1.30     5
#> 2 compact      20.1 3.39    28.3  3.78    47
#> 3 midsize      18.8 1.95    27.3  2.14    41
#> 4 minivan      15.8 1.83    22.4  2.06    11
#> 5 pickup       13   2.05    16.9  2.27    33
#> 6 subcompact   20.4 4.60    28.1  5.38    35
#> 7 suv          13.5 2.42    18.1  2.98    62
```

In performing summaries, it is possible to extract additional information; for example, which car has the maximum column value in a particular grouping. This can be achieved with the nth() function, which will return the nth value of an observation, and the which() function can be used to determine what the nth value should be. Here we can extract the model and manufacturer of those cars that have the maximum displacement.

```
mpg %>%
  group_by(class) %>%
  summarize(MaxDispl=max(displ),
            CarMax=dplyr::nth(model,which.max(displ)),
            ManuMax=dplyr::nth(manufacturer,which.max(displ)))
#> # A tibble: 7 x 4
#>   class     MaxDispl CarMax              ManuMax
#>   <chr>        <dbl> <chr>               <chr>
#> 1 2seater          7 corvette            chevrolet
#> 2 compact        3.3 camry solara        toyota
#> 3 midsize        5.3 grand prix          pontiac
#> 4 minivan          4 caravan 2wd         dodge
#> 5 pickup         5.9 ram 1500 pickup 4wd dodge
#> 6 subcompact     5.4 mustang             ford
#> 7 suv            6.5 k1500 tahoe 4wd     chevrolet
```

## 8.8 Additional `dplyr` functions

A number of additional `dplyr` functions are worth mentioning, as they can be used alongside the functions we have presented so far. These two functions are `case_when()` and `pull()`.

The function `case_when()` enables you to implement similar code to the `ifelse()` function. The arguments are a sequence of two-sided formulas, where the left-hand side (LHS) is used to formulate a condition, and the right-hand side (RHS) provides replacement values. The symbol ~ separates the two sides. For example, here we generate a random sample of ages, and then classify these according to `case_when()`.

```
x <- sample(1:99,5,replace = T)
x[5] <- 200

x
#> [1]  70  98   7   7 200

y <- dplyr::case_when(x < 18   ~ "Child",
                      x < 65   ~ "Adult",
                      x <= 99  ~ "Elderly",
                      TRUE     ~ "Unknown")
y
#> [1] "Elderly" "Elderly" "Child"   "Child"   "Unknown"
```

A couple of points worth noting. Once a LHS condition evaluates to TRUE, the value on the RHS of that condition is assigned to the output, and no other LHS conditions are evaluated. If no matches are found, the final statement TRUE is executed, and the output is then assigned to that value.

This function is often used with the `mutate()` function to create a new column based on the values in an existing column. For example, we can add a column to `mpg` based on whether the `cty` value is above or below the average value. Notice that we can use the tibble's column names within the body of `case_when()`.

```
mpg %>%
  dplyr::select(manufacturer:model,cty) %>%
  dplyr::mutate(cty_status=case_when(
    cty >= mean(cty) ~ "Above average",
    cty < mean(cty)  ~ "Below average",
    TRUE             ~ "Undefined"
  )) %>%
  dplyr::slice(1:5)
#> # A tibble: 5 x 4
#>   manufacturer model   cty cty_status
#>   <chr>        <chr> <int> <chr>
#> 1 audi         a4       18 Above average
#> 2 audi         a4       21 Above average
#> 3 audi         a4       20 Above average
#> 4 audi         a4       21 Above average
#> 5 audi         a4       16 Below average
```

A feature of the `dplyr` is that the functions we've explored always return a tibble. However, there may be cases when we need to return just one column, and while the `$` operator can be used, the function `pull()` is a better choice, especially when using pipes.

For example, if we wanted a vector that contained the unique values from the column `class` in `mpg`, we could write this as follows.

```
mpg %>%
  dplyr::pull(class) %>%
  unique()
#> [1] "compact"   "midsize"   "suv"        "2seater"
#> [5] "minivan"   "pickup"    "subcompact"
```

## 8.9   Mini-case: Summarizing total rainfall in 2017

In this mini-case, based on data stored in `aimsir17`, we use `dplyr` functions to summarize rainfall values, and `ggplot2` to visualize the results, with a focus on answering the following two questions.

1. Calculate the total annual rainfall for each weather station, and visualize using a bar chart.
2. Calculate the total monthly rainfall for two weather stations, "NEWPORT" and "DUBLIN AIRPORT", and visualize both using a time series graph.

For these tasks, we first need to load in the relevant packages, `dplyr`, `ggplot2`, and `aimsir17`. The `tibble` `observations` contains the weather data that we will summarize.

```
library(dplyr)
library(ggplot2)
library(aimsir17)
```

As a reminder, we can check the tibble `observations` to confirm the variables that are present.

```
observations
#> # A tibble: 219,000 x 12
#>    station  year month   day  hour date                  rain
#>    <chr>   <dbl> <dbl> <int> <int> <dttm>               <dbl>
#>  1 ATHENRY  2017     1     1     0 2017-01-01 00:00:00      0
#>  2 ATHENRY  2017     1     1     1 2017-01-01 01:00:00      0
#>  3 ATHENRY  2017     1     1     2 2017-01-01 02:00:00      0
#>  4 ATHENRY  2017     1     1     3 2017-01-01 03:00:00    0.1
#>  5 ATHENRY  2017     1     1     4 2017-01-01 04:00:00    0.1
#>  6 ATHENRY  2017     1     1     5 2017-01-01 05:00:00      0
#>  7 ATHENRY  2017     1     1     6 2017-01-01 06:00:00      0
#>  8 ATHENRY  2017     1     1     7 2017-01-01 07:00:00      0
#>  9 ATHENRY  2017     1     1     8 2017-01-01 08:00:00      0
#> 10 ATHENRY  2017     1     1     9 2017-01-01 09:00:00      0
#> # ... with 218,990 more rows, and 5 more variables: temp <dbl>,
#> #   rhum <dbl>, msl <dbl>, wdsp <dbl>, wddir <dbl>
```

The first task involves summing the variable `rain` for each of the weather stations. Therefore, it will involve grouping the entire tibble by the variable `station`, and performing a sum of `rain` for each of these groups. We will store this output in a tibble, and show the first ten rows.

```
annual_rain <- observations %>%
               dplyr::group_by(station) %>%
               dplyr::summarize(TotalRain=sum(rain,na.rm=T))
annual_rain
#> # A tibble: 25 x 2
#>    station          TotalRain
#>    <chr>                <dbl>
#>  1 ATHENRY              1199.
#>  2 BALLYHAISE            952.
#>  3 BELMULLET            1243.
#>  4 CASEMENT              705.
#>  5 CLAREMORRIS          1204.
#>  6 CORK AIRPORT         1162.
#>  7 DUBLIN AIRPORT        662.
#>  8 DUNSANY               810.
#>  9 FINNER               1222.
#> 10 GURTEEN               983.
#> # ... with 15 more rows
```

We can verify these aggregate values by performing a simple test based on the first weather station "ATHENRY".

```
test <- observations %>%
        dplyr::filter(station=="ATHENRY") %>%
        dplyr::pull(rain) %>%
        sum()
test
#> [1] 1199
annual_rain$TotalRain[1]
#> [1] 1199
```

Next, we can visualize the data (shown in Figure 8.1) from the tibble `annual_rain` to show the summaries for each station. Note that we add a horizontal line to show the maximum, and also that the x-axis labels are shown at an angle of 45 degrees. Note that we can use the functions `xlab()` and `ylab()` to conveniently set the titles for both axes.

```
ggplot(annual_rain,aes(x=station,y=TotalRain))+
  geom_bar(stat = "identity")+
  theme(axis.text.x=element_text(angle=45,hjust=1))+
  geom_hline(yintercept = max(annual_rain$TotalRain),color="red")+
  xlab("Weather Station") + ylab("Total Annual Rainfall")
```

Our second task also involves aggregating data, but in this case we focus on the monthly totals for two weather stations, "DUBLIN AIRPORT" and "NEWPORT". To set the scene, we can see first what the annual totals were

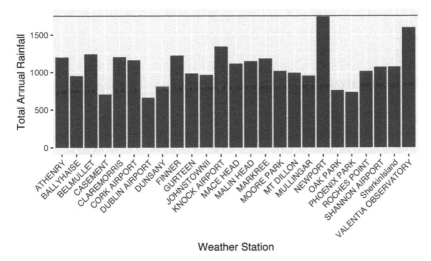

**FIGURE 8.1** Total annual rainfall for 2017 across 25 weather stations

for these two weather stations, and this clearly shows much more rain falls in Newport, located in Co. Mayo, and situated on Ireland's *Wild Atlantic Way*.

```
dplyr::filter(annual_rain,
        station %in% c("DUBLIN AIRPORT",
                        "NEWPORT"))
#> # A tibble: 2 x 2
#>    station         TotalRain
#>    <chr>             <dbl>
#> 1 DUBLIN AIRPORT      662.
#> 2 NEWPORT            1752.
```

In order to calculate the monthly values for both weather stations, we need to return to the `observations` tibble, filter all data for both stations, group by the variables `station` and `month`, and then apply the aggregation operation (i.e., sum of the total hourly rainfall).

```
monthly_rain <- observations %>%
                dplyr::filter(station %in% c("DUBLIN AIRPORT",
                                             "NEWPORT")) %>%
                dplyr::group_by(station, month) %>%
                dplyr::summarize(TotalRain=sum(rain,na.rm = T))
monthly_rain
#> # A tibble: 24 x 3
#> # Groups:   station [2]
#>     station         month TotalRain
#>     <chr>           <dbl>    <dbl>
#> 1 DUBLIN AIRPORT      1       22.8
```

```
#>  2 DUBLIN AIRPORT    2      41.6
#>  3 DUBLIN AIRPORT    3      67.2
#>  4 DUBLIN AIRPORT    4      10
#>  5 DUBLIN AIRPORT    5      43.5
#>  6 DUBLIN AIRPORT    6      86.4
#>  7 DUBLIN AIRPORT    7      42.2
#>  8 DUBLIN AIRPORT    8      73.2
#>  9 DUBLIN AIRPORT    9      82.3
#> 10 DUBLIN AIRPORT   10      47.8
#> # ... with 14 more rows
```

There are 24 observations in the summary tibble, which is to be expected, given that there are two stations, and 12 different months. We can also arrange the data by month (instead of station), yielding the following result.

```
dplyr::arrange(monthly_rain,month)
#> # A tibble: 24 x 3
#> # Groups:    station [2]
#>     station         month TotalRain
#>     <chr>           <dbl>    <dbl>
#>  1 DUBLIN AIRPORT      1     22.8
#>  2 NEWPORT             1     94.9
#>  3 DUBLIN AIRPORT      2     41.6
#>  4 NEWPORT             2    151
#>  5 DUBLIN AIRPORT      3     67.2
#>  6 NEWPORT             3    216.
#>  7 DUBLIN AIRPORT      4     10
#>  8 NEWPORT             4     31.2
#>  9 DUBLIN AIRPORT      5     43.5
#> 10 NEWPORT             5     64.2
#> # ... with 14 more rows
```

The total rainfall per month for each station, given that it is stored in the tibble `monthly_rain`, can now be visualized using `ggplot2`, using the functions `ggplot()`, geom_point(), and geom_line(). The function `xlim()`' placed limits on the x-axis values, and the plot is shown in Figure 8.2. Interestingly it shows that there was no month during 2017 in which the rainfall in Dublin Airport (east coast) was greater than that recorded in Newport (west coast).

```
ggplot(monthly_rain,aes(x=month,y=TotalRain,color=station))+
  geom_point()+geom_line()+
  theme(legend.position = "bottom")+
  scale_x_continuous(limits=c(1,12), breaks=seq(1,12))+
  labs(x="Month",
       y="Total Rain",
       title = "Monthly rainfall summaries calculated using dplyr")
```

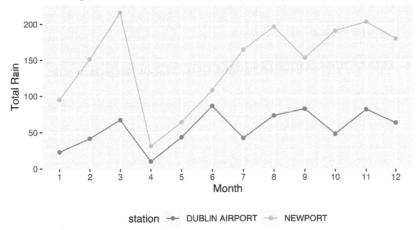

**FIGURE 8.2** Monthly summaries across two weather stations

## 8.10 Summary of R functions from Chapter 8

Functions from the package `dplyr` are now summarized. All functions are assumed to be part of `dplyr`, unless otherwise indicated.

| Function | Description |
|---|---|
| %>% | The tidyverse pipe operator (library magrittr). |
| %in% | Used to see if the left operand is in a vector. |
| filter() | Subsets rows based on column values. |
| slice() | Subsets rows based on their positions. |
| arrange() | Sorts rows based on column(s). |
| select() | Subsets columns based on names. |
| starts_with() | Matches starting column names. |
| ends_with() | Matches ending column names. |
| contains() | Matches column names that contain the input. |
| num_range() | Matches a numerical range. |
| matches() | Matches a regular expression. |
| mutate() | Creates new columns from existing variables. |
| group_by() | Converts a tibble to it into a grouped tibble. |
| ungroup() | Removes a tibble grouping. |
| summarize() | Creates a new data frame, based on summaries. |
| case_when() | Provides vectorization of multiple if_else() statement. |
| pull() | Similar to $ and useful if deployed with pipe |

## 8.11   Exercises

1. Based on the mpg dataset from ggplot2, generate the following tibble which filters all the cars with a cty value greater than the median. Ensure that your tibble contains the same columns, and with set.seed(100) sample five records using sample_n(), and store the result in the tibble ans.

```
ans
#> # A tibble: 5 x 7
#>    manufacturer model        displ  year   cty   hwy class
#>    <chr>        <chr>        <dbl> <int> <int> <int> <chr>
#> 1 nissan       maxima           3  1999    18    26 midsize
#> 2 volkswagen   gti              2  1999    19    26 compact
#> 3 subaru       impreza awd    2.5  2008    20    27 compact
#> 4 chevrolet    malibu         3.1  1999    18    26 midsize
#> 5 subaru       forester awd   2.5  2008    19    25 suv
```

2. Based on the aimsir17 tibble observations, generate the tibble jan which contains observations for two weather stations ("DUBLIN AIRPORT" and "MACE HEAD") during the month of January. Add a new column named WeatherStatus that contains three possible values: "Warning - Freezing" if the temperature is less than or equal to 0, "Warning - Very Cold" if the temperature is greater than zero and less than or equal to 4, and "No Warning" if the temparture is above 4. Make use of the case_when() function, and replicate the plot.

```
jan
#> # A tibble: 1,488 x 7
#>    station    month   day  hour date                 temp Weath~1
#>    <chr>      <dbl> <int> <int> <dttm>              <dbl> <chr>
#>  1 DUBLIN AI~     1     1     0 2017-01-01 00:00:00   5.3 No War~
#>  2 MACE HEAD      1     1     0 2017-01-01 00:00:00   5.6 No War~
#>  3 DUBLIN AI~     1     1     1 2017-01-01 01:00:00   4.9 No War~
#>  4 MACE HEAD      1     1     1 2017-01-01 01:00:00   5.4 No War~
#>  5 DUBLIN AI~     1     1     2 2017-01-01 02:00:00   5   No War~
#>  6 MACE HEAD      1     1     2 2017-01-01 02:00:00   4.7 No War~
#>  7 DUBLIN AI~     1     1     3 2017-01-01 03:00:00   4.2 No War~
#>  8 MACE HEAD      1     1     3 2017-01-01 03:00:00   4.7 No War~
#>  9 DUBLIN AI~     1     1     4 2017-01-01 04:00:00   3.6 Warnin~
#> 10 MACE HEAD      1     1     4 2017-01-01 04:00:00   4.5 No War~
#> # ... with 1,478 more rows, and abbreviated variable name
#> #   1: WeatherStatus
```

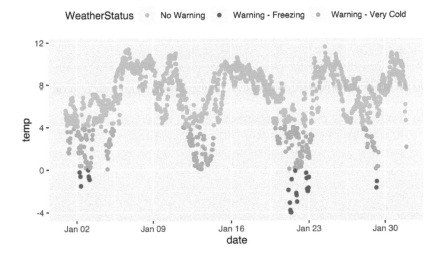

3. Generate the following tibble (`diam`) based on the `diamonds` tibble from `ggplot2`. Note that the column `PriceMaxColor` is the color of the diamond with the maximum price for a given cut.

```
diam
#> # A tibble: 5 x 5
#>    cut       NumberDiamonds CaratMean PriceMax PriceMaxColor
#>    <ord>              <int>     <dbl>    <int> <ord>
#> 1 Ideal              21551     0.703    18806 G
#> 2 Premium            13791     0.892    18823 I
#> 3 Very Good          12082     0.806    18818 G
#> 4 Good                4906     0.849    18788 G
#> 5 Fair                1610     1.05     18574 G
```

4. For each class of car, create the tibble `mpg1` that contains a new column that stores the rank of city miles per gallon (`cty`), from lowest to highest. Make use of the `rank()` function in R, and in this function call, set the argument `ties.method = "first"`.

```
mpg1
#> # A tibble: 234 x 7
#>   manufacturer model     displ  year   cty class   RankCty
#>   <chr>        <chr>     <dbl> <int> <int> <chr>     <int>
#> 1 chevrolet    corvette    5.7  1999    15 2seater       1
#> 2 chevrolet    corvette    6.2  2008    15 2seater       2
#> 3 chevrolet    corvette    7    2008    15 2seater       3
#> 4 chevrolet    corvette    5.7  1999    16 2seater       4
#> 5 chevrolet    corvette    6.2  2008    16 2seater       5
#> 6 audi         a4 quattro  2.8  1999    15 compact       1
```

```
#>  7 audi          a4 quattro   3.1  2008    15 compact      2
#>  8 audi          a4           2.8  1999    16 compact      3
#>  9 audi          a4 quattro   1.8  1999    16 compact      4
#> 10 volkswagen    jetta        2.8  1999    16 compact      5
#> # ... with 224 more rows
```

5.  Find the stations with the highest (`temp_high`) and lowest (`temp_low`) annual average temperature values. Use these variables to calculate the average monthly temperature values for the two stations (`m_temps`), and display the data in a plot.

```
temp_low
#> [1] "KNOCK AIRPORT"
temp_high
#> [1] "VALENTIA OBSERVATORY"
arrange(m_temps,month,station)
#> # A tibble: 24 x 3
#>     station              month AvrTemp
#>     <chr>                <dbl>   <dbl>
#>  1 KNOCK AIRPORT             1    5.18
#>  2 VALENTIA OBSERVATORY      1    8.03
#>  3 KNOCK AIRPORT             2    5.03
#>  4 VALENTIA OBSERVATORY      2    8.26
#>  5 KNOCK AIRPORT             3    6.78
#>  6 VALENTIA OBSERVATORY      3    9.36
#>  7 KNOCK AIRPORT             4    7.85
#>  8 VALENTIA OBSERVATORY      4    9.60
#>  9 KNOCK AIRPORT             5   11.6
#> 10 VALENTIA OBSERVATORY      5   12.6
#> # ... with 14 more rows
```

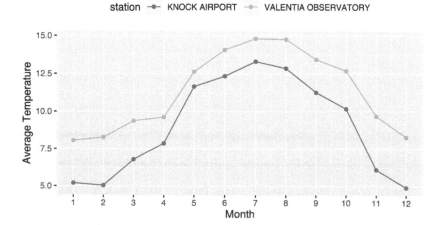

Monthly temperature summaries calculated using dplyr

# 9

---

# Relational Data with *dplyr* and *Tidying Data* with *tidyr*

---

It's rare that a data analysis involves only a single table of data. Typically you have many tables of data, and you must combine them to answer the questions you are interested in.

— Hadley Wickham and Garrett Grolemund (Wickham and Grolemund, 2016)

---

## 9.1  Introduction

The five main `dplyr` functions introduced in Chapter 8 are powerful ways to filter, transform, and summarize data stored in a tibble. However, when exploring data, there may be more than one tibble of interest, and these tibbles may share common information. For example, we may want to explore the correlation between flight delays and weather events, and these information sources may be contained in two different tibbles. In this chapter we show how information from more than one tibble can be merged, using what are known as *mutating joins* and *filtering joins*. Furthermore, we also explore a key idea known as *tidy data* and show how functions from the package `tidyr` can be used to streamline the data transformation pipeline.

Upon completing the chapter, you should understand the following:

- The main idea behind relational data, including concepts such as the primary key, foreign key, and composite key.
- The difference between mutating joins and filtering joins, and how `dplyr` provides functions for each type of join.

- The idea underlying tidy data, and how two functions from tidyr can be used to transform data from untidy to tidy format, and back again.
- How to solve all five test exercises.

### Chapter structure

- Section 9.2 describes the main ideas underlying the relational model.
- Section 9.3 introduces mutating joins, and how they can be achieved using inner_join(), left_join(), right_join(), and full_join().
- Section 9.4 summarizes filtering joins, and how to implement them using the functions semi_join() and anti_join().
- Section 9.5 presents the idea behind tidy data, and demonstrates the functions pivot_longer() and pivot_wider().
- Section 9.8 presents a study of how we can join two tables to explore relationships between wind speed and energy generated from wind.
- Section 9.9 provides a summary of the functions introduced in the chapter.
- Section 9.10 provides a number of short coding exercises to test your understanding of relational and tidy data.

## 9.2   Relational data

A relational model organizes data into one or more tables of columns and rows, where a unique key identifies each row. A key can be a single column value, known as a primary key, for example, a unique identifier for a person. For example, for the tibble aimsir17::stations, the primary key is the variable station, as this value is unique for each row, and therefore can be used to identify a single observation.

```
library(aimsir17)
library(dplyr)
stations %>% dplyr::filter(station=="MACE HEAD")
#> # A tibble: 1 x 5
#>    station    county height latitude longitude
#>    <chr>      <chr>   <dbl>    <dbl>     <dbl>
#> 1 MACE HEAD Galway      21     53.3     -9.90
```

A primary key from one table can also be a column in another table, and if this is the case, it is termed a *foreign key* in that table. For example, the tibble observations in the package aimsir17 contains the column station, which is not a primary key in this table, as it does not uniquely identify an observation. Therefore station is a foreign key in the table observations.

When multiple columns are used to uniquely identify an observation, this is termed a composite key. Many of the tibbles we deal with in this book have composite keys. For instance, in order to uniquely identify a row in observations, we need to use a combination of columns, for example, the variables year, station, month, day, and hour.

```
library(aimsir17)
library (dplyr)

observations %>% dplyr::filter(year==2017,
                               station=="MACE HEAD",
                               month==10,
                               day==16,
                               hour==12) %>%
                 dplyr::glimpse()
#> Rows: 1
#> Columns: 12
#> $ station <chr> "MACE HEAD"
#> $ year    <dbl> 2017
#> $ month   <dbl> 10
#> $ day     <int> 16
#> $ hour    <int> 12
#> $ date    <dttm> 2017-10-16 12:00:00
#> $ rain    <dbl> 2.9
#> $ temp    <dbl> 11.1
#> $ rhum    <dbl> 91
#> $ msl     <dbl> 976.6
#> $ wdsp    <dbl> 29
#> $ wddir   <dbl> 90
```

We will now provide a simple example, which will be used to show how two tables can be linked together via a foreign key. The idea here is that we can represent two important ideas from a company's information system, namely *products* and *orders*, in two separate, but related, tables. An overview of the structure is shown in Figure 9.1.

It shows two tables:

- **Products**, which has two variables, the *ProductID*, a primary key that uniquely identifies a product, and *Type*, which provides a category for each product.

- **Orders**, which contains three variables, the *OrderID* that uniquely identifies an order, the *Quantity*, which records the number of products ordered for each order, and the *ProductID*, which contains the actual product ordered. As *ProductID* is a primary key of the table **Products**, it is categorized as a foreign key in the table **Orders**.

| Primary key | |
|---|---|
| ProductID | Type |
| PR-1 | Computer |
| PR-2 | Tablet |
| PR-3 | Phone |
| PR-9 | Headphones |

| Primary key | | Foreign key |
|---|---|---|
| OrderID | Quantity | ProductID |
| OR-1 | 1 | PR-1 |
| OR-2 | 2 | PR-2 |
| OR-3 | 1 | PR-3 |
| OR-4 | 2 | PR-4 |
| OR-5 | 3 | PR-1 |

**FIGURE 9.1** An example of two tables with a relationship

A number of observations are worth noting for this example.

1. To simplify the example, we limit the number of product types for each order to one. Therefore, we have what is called a *one-to-many relationship*, where an order is for one product, but a product can be part of many orders.

2. There is an in-built inconsistency in the data, specifically, one of the orders ("OR-4") contains a product ("PR-4") that is not present in the product table. Normally, in a real-world database this would not be allowed, and through a process known as *referential integrity*, the database management system would enforce a constraint that a foreign key in one table would have to already be defined as a primary key in another table.

The reason we have added this inconsistency is that it helps us highlight the operation of joining functions used in dplyr. To prepare for the examples that follow, we recreate these tables in R using the tibble() function. First, we define the tibble products.

```
products <- tibble(ProductID=c("PR-1","PR-2","PR-3","PR-9"),
                   Type=c("Computer","Tablet","Phone","Headphones"))
products
#> # A tibble: 4 x 2
#>   ProductID Type
#>   <chr>     <chr>
#> 1 PR-1      Computer
#> 2 PR-2      Tablet
#> 3 PR-3      Phone
#> 4 PR-9      Headphones
```

We can see that the primary key `ProductID` uniquely identifies each product.

```
dplyr::filter(products,ProductID=="PR-9")
#> # A tibble: 1 x 2
#>    ProductID Type
#>    <chr>     <chr>
#> 1 PR-9       Headphones
```

Next we define the tibble `orders`, which also includes the foreign key column for the product associated with each order.

```
orders <- tibble(OrderID=c("OR-1","OR-2","OR-3","OR-4","OR-5"),
                 Quantity=c(1,2,1,2,3),
                 ProductID=c("PR-1","PR-2","PR-3","PR-4","PR-1"))
orders
#> # A tibble: 5 x 3
#>    OrderID Quantity ProductID
#>    <chr>      <dbl> <chr>
#> 1 OR-1           1 PR-1
#> 2 OR-2           2 PR-2
#> 3 OR-3           1 PR-3
#> 4 OR-4           2 PR-4
#> 5 OR-5           3 PR-1
```

We can confirm that `OrderID` uniquely identifies an order.

```
dplyr::filter(orders,OrderID=="OR-5")
#> # A tibble: 1 x 3
#>    OrderID Quantity ProductID
#>    <chr>      <dbl> <chr>
#> 1 OR-5           3 PR-1
```

We can now proceed to the core topics of this chapter, namely, exploring ways to combine data from different tables. These types of functions are called *joins*, and there are two main types, *mutating joins* and *filtering joins*.

---

## 9.3 Mutating joins

The first category of joins we explore is known as a *mutating join*, as these result in a tibble with additional columns. There are four functions that perform mutating joins: `inner_join()`, `left_join()`, `right_join()`, and `full_join()`.

### 9.3.1 `inner_join(x,y)`

The `inner_join()` function joins observations that appear in both tables, based on a common key, which need to be present in both tables. It takes the following arguments, and returns an object of the same type as x.

- x and y, a pair of tibbles or data frames to be joined.
- by, which is a character vector of variables to join by. In cases where the key column name is different, a named vector can be used, for example, by = c("key_x" = "key_y").

We can view the workings of `inner_join()` with our example.

```
i_j <- dplyr::inner_join(orders,products,by="ProductID")
i_j
#> # A tibble: 4 x 4
#>    OrderID Quantity ProductID Type
#>    <chr>       <dbl> <chr>     <chr>
#> 1 OR-1            1 PR-1      Computer
#> 2 OR-2            2 PR-2      Tablet
#> 3 OR-3            1 PR-3      Phone
#> 4 OR-5            3 PR-1      Computer
```

We can observe a number of points relating to the variable `i_j`.

- It adds the column `Type` to the tibble, containing information copied from the `products` tibble.
- It does not include any information on order "OR-4" from the orders tibble, and this is because "OR-4" is for product "PR-4", a product which does not appear in the `products` tibble.
- It also excludes information on product "PR-9", because there are no recorded orders for this product.

In summary, this mutating join is strict, as it only includes observations that have a common key across both tables.

### 9.3.2 `left_join(x,y)`

A left join will keep all observations in the tibble x, even if there is no match in tibble y. This is a widely used function, given that it maintains all the observations in x. We can now show two examples based on the tibbles `orders` and `products`.

```
l_j1 <- dplyr::left_join(orders,products,by="ProductID")
l_j1
#> # A tibble: 5 x 4
#>    OrderID Quantity ProductID Type
#>    <chr>       <dbl> <chr>     <chr>
#> 1 OR-1            1 PR-1      Computer
```

```
#> 2 OR-2           2 PR-2        Tablet
#> 3 OR-3           1 PR-3        Phone
#> 4 OR-4           2 PR-4        <NA>
#> 5 OR-5           3 PR-1        Computer
```

All of the orders are included by default, and if there is no matching key in the paired tibble, a value of NA is returned. Here we can see that there is no product with an identifier of PR-4 in the products table.

This can be further clarified if the argument keep=TRUE is passed into the join function, as this will maintain both keys in the output. Interestingly, this output clearly shows that the key on tibble y is missing, and note that each table key is further annotated with .x for the first tibble, and .y for the second tibble.

```
l_j2 <- dplyr::left_join(orders,products,by="ProductID",keep=TRUE)
l_j2
#> # A tibble: 5 x 5
#>    OrderID Quantity ProductID.x ProductID.y Type
#>    <chr>      <dbl> <chr>       <chr>       <chr>
#> 1 OR-1           1 PR-1        PR-1        Computer
#> 2 OR-2           2 PR-2        PR-2        Tablet
#> 3 OR-3           1 PR-3        PR-3        Phone
#> 4 OR-4           2 PR-4        <NA>        <NA>
#> 5 OR-5           3 PR-1        PR-1        Computer
```

We can now show the use of left join where the first two arguments are reversed.

```
l_j3 <- dplyr::left_join(products,orders,by="ProductID")
l_j3
#> # A tibble: 5 x 4
#>    ProductID Type       OrderID Quantity
#>    <chr>     <chr>      <chr>      <dbl>
#> 1 PR-1      Computer   OR-1           1
#> 2 PR-1      Computer   OR-5           3
#> 3 PR-2      Tablet     OR-2           2
#> 4 PR-3      Phone      OR-3           1
#> 5 PR-9      Headphones <NA>          NA
```

Notice that all the observations in x are maintained (the four products), and matches in y are included. Because product PR-1 appears in two orders, two of the rows contain PR-1. We can also see that PR-9 has no OrderID or Quantity, as this product is not contained in the orders tibble. Again, we can also see the tibble with the product keys from both tibbles x and y maintained.

```
l_j4 <- dplyr::left_join(products,orders,by="ProductID",keep=TRUE)
l_j4
#> # A tibble: 5 x 5
```

```
#>     ProductID.x Type         OrderID Quantity ProductID.y
#>     <chr>       <chr>        <chr>      <dbl> <chr>
#> 1 PR-1          Computer     OR-1           1 PR-1
#> 2 PR-1          Computer     OR-5           3 PR-1
#> 3 PR-2          Tablet       OR-2           2 PR-2
#> 4 PR-3          Phone        OR-3           1 PR-3
#> 5 PR-9          Headphones   <NA>          NA <NA>
```

### 9.3.3  `right_join(x,y)`

A right join keeps all observations in the tibble y. In this example, we can see
that all the product information is shown (there are five products in all), but
that the order "OR-4" is missing, as that is for "PR-4", which is not present
in the products table.

```
r_j1 <- dplyr::right_join(orders,products,by="ProductID")
r_j1
#> # A tibble: 5 x 4
#>   OrderID Quantity ProductID Type
#>   <chr>      <dbl> <chr>     <chr>
#> 1 OR-1           1 PR-1      Computer
#> 2 OR-2           2 PR-2      Tablet
#> 3 OR-3           1 PR-3      Phone
#> 4 OR-5           3 PR-1      Computer
#> 5 <NA>          NA PR-9      Headphones
```

We can also perform the right join where the first tibble is products. Again,
with the right join all observations in the y tibble are returned, and therefore
the product "PR-9" is missing, as it is not linked to any order.

```
r_j2 <- dplyr::right_join(products,orders,by="ProductID")
r_j2
#> # A tibble: 5 x 4
#>   ProductID Type       OrderID Quantity
#>   <chr>     <chr>      <chr>      <dbl>
#> 1 PR-1      Computer   OR-1           1
#> 2 PR-1      Computer   OR-5           3
#> 3 PR-2      Tablet     OR-2           2
#> 4 PR-3      Phone      OR-3           1
#> 5 PR-4      <NA>       OR-4           2
```

### 9.3.4 `full_join(x,y)`

A full join keeps all observations in both x and y. The same overall result is obtained regardless of which tibble is the first one. For example, here is a full join of products and orders.

```
f_j1 <- dplyr::full_join(products,orders,by="ProductID")
f_j1
#> # A tibble: 6 x 4
#>   ProductID Type       OrderID Quantity
#>   <chr>     <chr>      <chr>      <dbl>
#> 1 PR-1      Computer   OR-1           1
#> 2 PR-1      Computer   OR-5           3
#> 3 PR-2      Tablet     OR-2           2
#> 4 PR-3      Phone      OR-3           1
#> 5 PR-9      Headphones <NA>          NA
#> 6 PR-4      <NA>       OR-4           2
```

We can also see the result of starting the full join operation with `orders`.

```
f_j2 <- dplyr::full_join(orders,products,by="ProductID")
f_j2
#> # A tibble: 6 x 4
#>   OrderID Quantity ProductID Type
#>   <chr>      <dbl> <chr>     <chr>
#> 1 OR-1           1 PR-1      Computer
#> 2 OR-2           2 PR-2      Tablet
#> 3 OR-3           1 PR-3      Phone
#> 4 OR-4           2 PR-4      <NA>
#> 5 OR-5           3 PR-1      Computer
#> 6 <NA>          NA PR-9      Headphones
```

## 9.4 Filtering joins

There are another set of joins that can be used to filter observations in tibble x, based on their relationship with values in another table. As the name suggests, these are *filtering joins*, and perform a similar task to the regular `filter()` function, except that information from two tables is used.

### 9.4.1 `semi_join(x,y)`

This function will keep all the observations in x that have a matching column in y. In our manufacturing example, we can perform this join based on the `ProductID`, starting with the orders table.

```
s_j1 <- dplyr::semi_join(orders,products,by="ProductID")
s_j1
#> # A tibble: 4 x 3
#>   OrderID Quantity ProductID
#>   <chr>      <dbl> <chr>
#> 1 OR-1           1 PR-1
#> 2 OR-2           2 PR-2
#> 3 OR-3           1 PR-3
#> 4 OR-5           3 PR-1
```

Notice that this join only presents observations from the orders tibble, and this
is restricted to products that are present in the products tibble. The semi-join
can also be performed starting with the products table, and so it will only
show those products that are linked to an order present in the orders table.

```
s_j2 <- dplyr::semi_join(products,orders,by="ProductID")
s_j2
#> # A tibble: 3 x 2
#>   ProductID Type
#>   <chr>     <chr>
#> 1 PR-1      Computer
#> 2 PR-2      Tablet
#> 3 PR-3      Phone
```

Here we observe that "PR-9" is missing, as it is not present in the orders table.

### 9.4.2   anti_join(x,y)

This filtering function will keep all the observations in x that do have a matching
column in y. Again, this is a filtering join, and therefore only observations from
the first tibble are returned. The function can be applied to our example, and
yields the following results.

```
a_j1 <- dplyr::anti_join(orders,products,by="ProductID")
a_j1
#> # A tibble: 1 x 3
#>   OrderID Quantity ProductID
#>   <chr>      <dbl> <chr>
#> 1 OR-4           2 PR-4
```

This results confirms that, in the orders tibble, the sole product that is not
represented in products is "PR-4". We can also explore the result for products
to see which product is not linked to an order.

```
a_j2 <- dplyr::anti_join(products,orders,by="ProductID")
a_j2
#> # A tibble: 1 x 2
```

```
#>    ProductID Type
#>    <chr>     <chr>
#> 1 PR-9       Headphones
```

The result discovers that "PR-9" is the only product not linked to an order.

## 9.5 Tidy data

When working with datasets, a key aim is to structure the dataset so as to facilitate analysis, and the output of this structuring process is a *tidy dataset*. We often use the term *rectangular data* to describe a dataset in R, where we have rows and columns, similar to a spreadsheet. However, not all rectangular data is in tidy data format, because a tidy dataset (Wickham, 2014) is defined as one where:

1. Each variable is a column,
2. Each observation is a row, and
3. Each type of observational unit is a table.

In order to explore this idea, consider Figure 9.2, which shows two tables, each of which stores the same information in different rectangular formats. The tables contain synthetic information for the exam results of two students ("ID1" and "ID2"), across three subjects ("CX101","CX102", and "CX103").

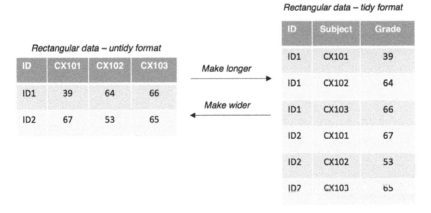

**FIGURE 9.2** Two ways for storing the same data

Even though the same information is present in both tables, we can observe that the structure of the tables differs, and this has an impact on how we can analyze the data using the tools of the tidyverse. We term the format on the left *untidy data*, because *each column is not a variable*. Specifically, the columns "CX101", "CX102", and "CX103" are subject names, and therefore are what we call instances of the variable "Subject". Also, in this case, each row contains three observations of the same variable.

If we could find a way to transform the table on the left to that on the right, we would now have a tidy data format. In this table, there are now three variables: *ID*, *Subject*, and *Grade*, and notice that all of the grades are present, but that each column is a variable, and each row is an observation.

That is not to say that the untidy format is not useful, as it displays the data in a more readable format, for example, to show on a report or a presentation. However, for the process of data analysis using the tidyverse, the format on the right is much more effective, as it supports direct manipulation using the tools provided by `ggplot2` and `dplyr`.

We will now re-create the untidy data in R using atomic vectors and the `tibble()` function. We include the library `tidyr`, as this contains functions to transform the data to tidy format.

```r
library(dplyr)
library(tidyr)
set.seed(100)
N = 2


res <- tibble(ID=paste0("ID",1:N),
              CX101=sample(30:70,N),
              CX102=sample(40:80,N),
              CX103=sample(60:70,N))
res
#> # A tibble: 2 x 4
#>    ID     CX101 CX102 CX103
#>    <chr> <int> <int> <int>
#> 1 ID1      39    64    66
#> 2 ID2      67    53    65
```

Given that the variable `res` represents the data in untidy format, we will now explore how this can be transformed using the `tidyr` function `pivot_longer()`.

---

## 9.6   Making data longer using `pivot_longer()`

The `tidyr` function `pivot_longer()` is designed to create this new tibble, and it accepts the following main arguments.

- data, the tibble on which to perform the pivot action on.
- cols, the columns to pivot into longer format.
- names_to, a character vector specifying the new column to create, and the values of this column will be the column names specified in the argument cols.
- values_to, a character vector specifying the column name where data will be stored.

With these arguments, we can now convert the tibble to a longer format.

```
res_l <- tidyr::pivot_longer(res,
                             `CX101`:`CX103`,
                             names_to="Subject",
                             values_to="Grade")
```

```
res_l
#> # A tibble: 6 x 3
#>    ID     Subject Grade
#>    <chr>  <chr>   <int>
#> 1 ID1     CX101      39
#> 2 ID1     CX102      64
#> 3 ID1     CX103      66
#> 4 ID2     CX101      67
#> 5 ID2     CX102      53
#> 6 ID2     CX103      65
```

The advantage of this transformation to tidy data can be observed when using other dplyr functions such as summarize(). For example, it is now straightforward to process the new tibble to calculate the average, minimum, and maximum grade for each student.

```
res_l %>%
   dplyr::group_by(ID) %>%
   dplyr::summarize(AvrGrade=mean(Grade),
                    MinGrade=min(Grade),
                    MaxGrade=max(Grade))
#> # A tibble: 2 x 4
#>    ID   AvrGrade MinGrade MaxGrade
#>    <chr>   <dbl>    <int>    <int>
#> 1 ID1      56.3       39       66
#> 2 ID2      61.7       53       67
```

Furthermore, because it is in tidy data format, we can use ggplot2 to visualize the student grades, and this is shown in Figure 9.3.

```
ggplot(res_l,aes(x=ID,y=Grade,fill=Subject ))+
   geom_bar(stat = "identity", position="dodge")
```

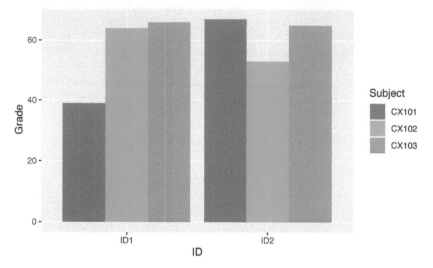

**FIGURE 9.3** Showing grades for two students

---

## 9.7    Making data wider using `pivot_wider()`

There are times when you may need to perform the inverse to `pivot_longer()`, namely, converting data from a long format to a wider format. This can be effective for presenting results in a wider tabular format, and the function `pivot_wider()` can be used to support this transformation. The arguments for `pivot_wider()` include:

- `data`, the data frame to transform.
- `names_from` and `values_from`, two arguments describing which column to get the column names from, and which column to extract the cell values from.

We can revisit the grades example, and convert the longer tibble (`res_l`) to a wider tibble (`res_w`) as follows.

```
res_w <- res_l %>%
          tidyr::pivot_wider(names_from=Subject,
                             values_from = Grade)
res_w
#> # A tibble: 2 x 4
#>    ID    CX101 CX102 CX103
#>    <chr> <int> <int> <int>
#> 1 ID1      39    64    66
#> 2 ID2      67    53    65
```

Note that this is the same as the original wider tibble we generated (`res`). It is worth noting that more complex operations can be performed on both `pivot_longer()` and `pivot_wider()`, and it is worth checking out the options by viewing the function documentation from `tidyr`, for example, `?pivot_longer`.

---

## 9.8 Mini-case: Exploring correlations with wind energy generation

The aim of this example is to show how to combine related tibbles using `dplyr`, in order to explore new relationships. We use existing data from the package `aimsir17`, and in particular two tibbles:

- `observations`, which contains hourly information from 2017 for 25 weather stations across Ireland, including the rainfall, temperature, humidity, mean sea level atmospheric pressure, wind speed, and wind direction.

- `eirgrid17`, which holds information, recorded four times per hour, on energy demand and supply from the Irish Grid in 2017, including overall demand, energy generated, and wind energy generated.

Our goal is to investigate, for each weather station, the correlation between the wind speed recorded, and the national wind energy generated. It is based on the simplified assumption that wind speed is transformed quickly into power generated; therefore, we would anticipate to see a positive correlation between wind speed and wind energy generated.

Our approach is summarized in Figure 9.4 which shows two existing tibbles (`eirgrid17` and `observations`), and two new tibbles (`eirgrid17_h` and `weather_energy`). The overall steps are as follows:

- The two tibbles we are seeking to join share common variables related to the observation times. For example, `eirgrid17` and `observations` both have columns for `year`, `month`, `day`, and `hour`.

- However, the electricity grid data contained in `eirgrid17` further divides an hour into four equal observations of 15 minutes each. In order to ensure that this data is fully aligned with the weather observations, we use `dplyr` to get the mean wind energy generated for every hour. This information is stored in the tibble `eirgrid17h`.

- Following this, we perform a mutating join between the tibbles `eirgrid17_h` and `observations` and this generates a new tibble, `weather_energy`. This tibble will contain the data needed for the correlation analysis.

We can now explore the code to perform this analysis. First, we include the relevant libraries.

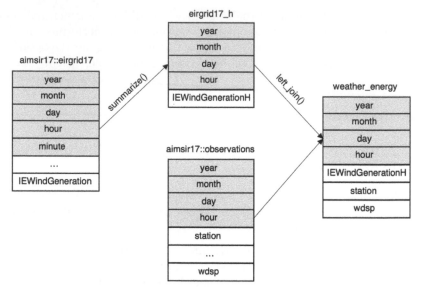

**FIGURE 9.4** Creating a tibble for analysis using a mutating join

```
library(aimsir17)
library(dplyr)
library(ggplot2)
```

A quick check of the tibble `eirgrid17` confirms the available data, and this includes our main target variable `IEWindGeneration`.

```
dplyr::glimpse(eirgrid17)
#> Rows: 35,040
#> Columns: 15
#> $ year              <dbl> 2017, 2017, 2017, 2017, 2017, 2017, ~
#> $ month             <dbl> 1, 1, 1, 1, 1, 1, 1, 1, 1, 1, 1, 1, ~
#> $ day               <int> 1, 1, 1, 1, 1, 1, 1, 1, 1, 1, 1, 1, ~
#> $ hour              <int> 0, 0, 0, 0, 1, 1, 1, 1, 2, 2, 2, 2, ~
#> $ minute            <int> 0, 15, 30, 45, 0, 15, 30, 45, 0, 15,~
#> $ date              <dttm> 2017-01-01 00:00:00, 2017-01-01 00:~
#> $ NIGeneration      <dbl> 889.0, 922.2, 908.1, 918.8, 882.4, 8~
#> $ NIDemand          <dbl> 775.9, 770.2, 761.2, 742.7, 749.2, 7~
#> $ NIWindAvailability <dbl> 175.1, 182.9, 169.8, 167.5, 174.1, 1~
#> $ NIWindGeneration  <dbl> 198.2, 207.8, 193.1, 190.8, 195.8, 2~
#> $ IEGeneration      <dbl> 3289, 3282, 3224, 3171, 3190, 3185, ~
#> $ IEDemand          <dbl> 2921, 2884, 2806, 2719, 2683, 2650, ~
#> $ IEWindAvailability <dbl> 1064.8, 965.6, 915.4, 895.4, 1028.0,~
```

```
#> $ IEWindGeneration    <dbl> 1044.7, 957.7, 900.5, 870.8, 998.3, ~
#> $ SNSP                <chr> "28.4%", "26.4%", "25.2%", "24.7%", ~
```

Next, we create the tibble `eirgrid17_h`, which gets the mean wind power
generated for each hour, where this operation is completed on each of four
hourly observations. Note that we also ungroup the tibble following the call to
`summarize()`.

```
eirgrid17_h <- eirgrid17 %>%
                dplyr::group_by(year,month,day,hour) %>%
                dplyr::summarize(IEWindGenerationH=
                                mean(IEWindGeneration,
                                     na.rm=T)) %>%
                dplyr::ungroup()
dplyr::slice(eirgrid17_h,1:4)
#> # A tibble: 4 x 5
#>     year month   day  hour IEWindGenerationH
#>    <dbl> <dbl> <int> <int>            <dbl>
#> 1   2017     1     1     0             943.
#> 2   2017     1     1     1            1085.
#> 3   2017     1     1     2            1284.
#> 4   2017     1     1     3            1254.
```

It is often helpful to visualize elements of the dataset, and here we explore the
dataset for the average hourly wind energy generated for new year's day in
2017. This is shown in Figure 9.5. We can see that the summarize function
has collapsed the four hourly values into a single mean observation.

```
ggplot(dplyr::filter(eirgrid17_h,month==1,day == 1),
       aes(x=hour,y=IEWindGenerationH))+
       geom_point()+
       geom_line()
```

With the alignment now complete for four of the variables (year, month, day,
and hour), we use a mutating join to merge both tables. In this case, we want
to maintain all of the energy observations; therefore, a left join from `eirgrid_h`
is used. The target tibble is a filtered form of observations, where we have
removed all records with missing wind speed values. For example, a number of
weather stations have no observations for wind speed, as can be seen from the
following code. (Here we introduce the `dplyr` function `count()`, which allows
you to quickly count the unique values in one or more variables).

```
observations %>%
  dplyr::filter(is.na(wdsp)) %>%
  dplyr::group_by(station) %>%
  dplyr::count()
#> # A tibble: 5 x 2
```

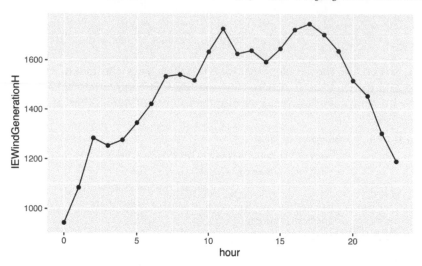

**FIGURE 9.5** Wind energy generated for January 1st

```
#> # Groups:    station [5]
#>   station                   n
#>   <chr>                 <int>
#> 1 CLAREMORRIS               5
#> 2 MARKREE                8760
#> 3 PHOENIX PARK           8760
#> 4 ROCHES POINT              1
#> 5 VALENTIA OBSERVATORY     21
```

The exclusion of missing values (stored in `obs1`) and the join operation (stored in `weather_energy`) are now shown below. To simplify the presentation, only those variables needed for the correlation analysis (`wdsp` and `IEWindGeneration`) are retained.

```
obs1 <- observations %>%
       dplyr::filter(!is.na(wdsp))


weather_energy <- dplyr::left_join(eirgrid17_h,
                                   obs1,
                                   by=c("year",
                                        "month",
                                        "day",
                                        "hour")) %>%
                dplyr::select(year,month,day,hour,
                              IEWindGenerationH,
                              station,
                              wdsp)
```

```
weather_energy
#> # A tibble: 201,430 x 7
#>     year month   day  hour IEWindGenerationH station         wdsp
#>    <dbl> <dbl> <int> <int>             <dbl> <chr>          <dbl>
#>  1  2017     1     1     0              943. ATHENRY            8
#>  2  2017     1     1     0              943. BALLYHAISE         5
#>  3  2017     1     1     0              943. BELMULLET         13
#>  4  2017     1     1     0              943. CASEMENT           8
#>  5  2017     1     1     0              943. CLAREMORRIS        8
#>  6  2017     1     1     0              943. CORK AIRPORT      11
#>  7  2017     1     1     0              943. DUBLIN AIRPORT    12
#>  8  2017     1     1     0              943. DUNSANY            6
#>  9  2017     1     1     0              943. FINNER            12
#> 10  2017     1     1     0              943. GURTEEN            7
#> # ... with 201,420 more rows
```

With the new table created that will form the basis of our analysis, we can
visually explore the relationship, as can be seen in Figure 9.6. Because there
are many points in the full dataset, to simplify the visualization we sample a
number of points from `weather_energy`, and show this relationship, including
a visualization of the linear model, and limit the analysis to just one station
(*Mace Head*). The function `geom_jitter()` will display overlapping values by
"shaking" the points so that an indication of the number of data points in the
general area of the graph is provided.

```
set.seed(100)
obs_sample <- weather_energy %>%
              dplyr::filter(station %in% c("MACE HEAD")) %>%
              dplyr::sample_n(300)

ggplot(obs_sample,aes(x=wdsp,y=IEWindGenerationH))+
  geom_point()+
  geom_jitter()+
  geom_smooth(method="lm")
```

The line does show the basis of a positive linear relationship between the
variables, where an increase in wind speed is associated with an increase in
wind energy generation. We can now analyze the strength of this correlation.
For this task, we can group the observations by station and then call the R
function `cor()` to calculate the correlation coefficient. Note that the default
method used is the Pearson method, and the calculated values are displayed
in descending order.

```
corr_sum <- weather_energy %>%
            dplyr::group_by(station) %>%
            dplyr::summarize(Correlation=cor(wdsp,
```

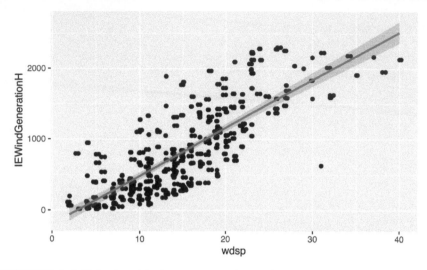

**FIGURE 9.6** Relationship between wind speed and wind energy generated

```
                                              IEWindGenerationH)) %>%
                dplyr::arrange(desc(Correlation))
corr_sum
#> # A tibble: 23 x 2
#>    station              Correlation
#>    <chr>                    <dbl>
#>  1 GURTEEN                  0.806
#>  2 KNOCK AIRPORT            0.802
#>  3 MACE HEAD                0.800
#>  4 VALENTIA OBSERVATORY     0.778
#>  5 SHANNON AIRPORT          0.778
#>  6 MULLINGAR                0.771
#>  7 SherkinIsland            0.768
#>  8 CORK AIRPORT             0.768
#>  9 ROCHES POINT             0.767
#> 10 ATHENRY                  0.764
#> # ... with 13 more rows
```

We can display the results in a bar chart shown in Figure 9.7. Note that stat="identity" is used as the bar chart data is available. Interestingly, for this sample, a strong association is shown between the two variables across all the weather stations.

```
ggplot(corr_sum,aes(x=Correlation,y=station))+
  geom_bar(stat="identity")
```

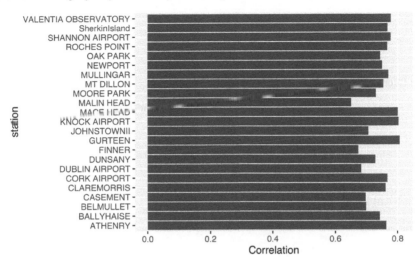

**FIGURE 9.7** A comparison of the correlation values for each weather station

## 9.9 Summary of R functions from Chapter 9

The most important relational functions from the package `dplyr` are now summarized, along with two functions from the package `tidyr`.

| Function | Description |
| --- | --- |
| count() | Counts the unique values in one or more variables. |
| inner_join(x,y) | Maintains observations that appear in both x and y. |
| left_join(x,y) | Keeps all observations in x, and joins with y. |
| right_join(x,y) | Keeps all observations in y, and joins with x. |
| full_join(x,y) | Keeps all observations in x and y. |
| semi_join(x,y) | Keeps all rows from x that have a match in y. |
| anti_join(x,y) | Keeps all rows from x that have no match in y. |
| pivot_longer() | Tidy data operation to lengthen data. |
| pivot_wider() | Inverse of pivot_longer(). |

## 9.10 Exercises

1. Based on the package `nycflights13`, which can be downloaded from CRAN, generate the following tibble based on the first three records from the tibble `flights`, and the airline name from `airlines`.

```
first_3a
#> # A tibble: 3 x 5
#>   time_hour               origin dest  carrier name
#>   <dttm>                  <chr>  <chr> <chr>   <chr>
#> 1 2013-01-01 05:00:00 EWR        IAH   UA      United Air Lines Inc.
#> 2 2013-01-01 05:00:00 LGA        IAH   UA      United Air Lines Inc.
#> 3 2013-01-01 05:00:00 JFK        MIA   AA      American Airlines Inc.
```

2. Based on the package nycflights13, generate the following tibble based on the first three records from the tibble flights, and the airport details from the tibble airports. The result should show the full name of the destination airport.

```
first_3b
#> # A tibble: 3 x 5
#>   time_hour               origin dest  name                     tzone
#>   <dttm>                  <chr>  <chr> <chr>                    <chr>
#> 1 2013-01-01 05:00:00 EWR        IAH   George Bush Intercontin~ Amer~
#> 2 2013-01-01 05:00:00 LGA        IAH   George Bush Intercontin~ Amer~
#> 3 2013-01-01 05:00:00 JFK        MIA   Miami Intl               Amer~
```

3. Using an appropriate filtering join, find all the destinations in flights that do not have an faa code in airports. The output should be a vector.

```
dest_not_in_airports
#> [1] "BQN" "SJU" "STT" "PSE"
```

4. Using the tools of tidyr, convert the table tidyr::table4b to the following tibble.

```
t4b_long
#> # A tibble: 6 x 3
#>   country     Year  Population
#>   <chr>       <chr>      <dbl>
#> 1 Afghanistan 1999    19987071
#> 2 Afghanistan 2000    20595360
#> 3 Brazil      1999   172006362
#> 4 Brazil      2000   174504898
#> 5 China       1999  1272915272
#> 6 China       2000  1280428583
```

5. For three weather stations, "MACE HEAD", "DUBLIN AIRPORT" and "NEWPORT", use the tools of dplyr and tidyr to generate

the following informative tibble summarizing the rainfall for each quarter of the year. Note that the final operation for the data analysis pipeline will be a call to the `tidyr` function `pivot_wider`.

```
sum_wide
#> # A tibble: 3 x 5
#>   station           Q1    Q2    Q3    Q4
#>   <chr>          <dbl> <dbl> <dbl> <dbl>
#> 1 DUBLIN AIRPORT  132.  140.  198.  192.
#> 2 MACE HEAD       266.  164.  308.  376.
#> 3 NEWPORT         462.  204.  513.  573.
```

# 10

## Processing Data with *purrr*

> R has numerous ways to iterate over elements of a list (or vector), and Hadley Wickham aimed to improve on and standardise that experience with the purrr package.
>
> — Jared P. Lander (Lander, 2017)

## 10.1 Introduction

Earlier in Chapter 4.6 we introduced functionals in R, and specifically, the functional lapply() which was used to iterate over an atomic vector, list, or data frame. The tidyverse package purrr provides a comprehensive set of functions that can be used to iterate over data structures, and also integrate with other elements of the tidyverse, for example, the package dplyr. This chapter introduces functions from purrr, and upon completion, you should understand:

- The function map(.x,.f), how it can be used to iterate over lists, its similarities and differences when compared to lapply().
- The syntax for function shortcuts that can be used in map().
- The specialized versions of map_*() which, instead of returning a list, will return a specified data type.
- The utility of using dplyr::group split() with map() to process groups of data stored in tibbles.
- The advantage of using tidyr::nest() with map() to support a statistical modelling workflow.
- The key functions that allow you to manipulate lists.
- How to solve all five test exercises.

## Chapter structure

- Section 10.2 introduces `map()`, and shows how it can be used to process lists.
- Section 10.3 explores additional map functions that generate a range of outputs, including atomic vectors and data frames.
- Section 10.4 shows how the functions `map2()` and `pmap()` can be used to process two and more inputs.
- Section 10.5 shows how `purrr` can be integrated with functions from `dplyr` and `tidyr`, and in doing so integrate tibbles with `purrr` to create a data processing pipeline.
- Section 10.6 summarizes additional functions that can be used to process lists.
- Section 10.7 summarizes a mini-case that highlights how `purrr` can be used to support a statistical modelling workflow.
- Section 10.8 provides a summary of all the functions introduced in the chapter.
- Section 10.9 provides a number of short coding exercises to test your understanding of `purrr`.

## 10.2    Iteration using `map()`

The idea of the `map()` function is to provide a mechanism to iterate over an input list or a vector. It applies a function to each input element, and returns the result within a list that is exactly the same length as the input. We call these functions *functionals*, as they accept a function as an argument, and use that function in order to generate the output.

This is similar to the `lapply()` function introduced in Chapter 4, although it does have a number of additional features and also works well within the tidyverse. An high-level overview of `map()` is shown in Figure 10.1 and shows the idea that map takes in a vector (`.x`) and a function (`.f`), and it generates an output list that contains the result of the function applied to each list element. This list will be the same size as the input list `.x`.

The general format of this function is `map(.x, .f)`, where:

- `.x` is a list, or an atomic vector. If .x has named elements, the return value will preserve those names.
- `.f` can be a function, formula, or vector.

We will explore each variation of `.f` in turn, as there are three formats that can be used:

- First, an anonymous function can be used with `map()`, just as was used for the `lapply()` function. Here, we define an anonymous function that finds the

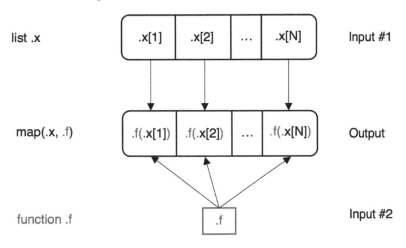

**FIGURE 10.1** The map function in purrr

square of an input atomic vector. Note that the length of the output list is the same as the length of the input atomic vector.

```
library(purrr)

o1 <- purrr::map(c(1,2,3,2),function(x)x^2)
str(o1)
#> List of 4
#>  $ : num 1
#>  $ : num 4
#>  $ : num 9
#>  $ : num 4
```

- Second, a *formula* (which is defined by ~) can be used as a function argument, and it is essentially a function shortcut in purrr, and commonly used in map(). We can use the formula mechanism to rewrite the first anonymous function as a shortcut, where ~ marks the start of the function, and .x as the function parameter, where . can also be used to represent the function parameter.

```
o2 <- purrr::map(c(1,2,3,2),~.x^2)
str(o2)
#> List of 4
#>  $ : num 1
#>  $ : num 4
#>  $ : num 9
#>  $ : num 4
o3 <- purrr::map(c(1,2,3,2),~.^2)
str(o3)
```

```
#> List of 4
#>  $ : num 1
#>  $ : num 4
#>  $ : num 9
#>  $ : num 4
```

We can see how `purrr` deals with formulas by calling the utility function `purrr::as_mapper()` with the formula as a parameter.

```
purrr::as_mapper(~.x^2)
#> <lambda>
#> function (..., .x = ..1, .y = ..2, . = ..1)
#> .x^2
#> attr(,"class")
#> [1] "rlang_lambda_function" "function"
```

As can be seen, this wrapper creates a full R function, and it captures the function logic with the command `.x^2`. The first argument can be accessed using `.x` or `.`, and a second parameter, which is used in functions such as `map2()` is accessed using `.y`. For more arguments, they can be accessed using `..1`, `..2`, `..3`, etc. Overall, it is optional for you to use either a function or a formula with `map()`, and normally for short functions the formula approach is used.

- Third, the map functions can be used to select a list element by name, and these are beneficial when working with lists that are deeply nested (Wickham, 2019). For example, in the `sw_films` list we explored earlier, passing in the name of a list element will result in the extraction of the associated data, in this case, the names of all the directors.

```
library(repurrrsive)
dirs <- sw_films %>% purrr::map("director") %>% unique()
str(dirs)
#> List of 4
#>  $ : chr "George Lucas"
#>  $ : chr "Richard Marquand"
#>  $ : chr "Irvin Kershner"
#>  $ : chr "J. J. Abrams"
```

## 10.3   Additional `map_*` functions

While `map()` will always return a list, there may be circumstances where different formats are required, for example, an atomic vector. To address this, `purrr` provides a set of additional functions that specify the result type. These include:

- `map_dbl()`, which returns an atomic vector of type `double`.
- `map_chr()`, which returns an atomic vector of type `character`.
- `map_lgl()`, which returns an atomic vector of type `logical`.
- `map_int()`, which returns an atomic vector of type `integer`.
- `map_df()`, which returns a data frame or tibble.

Each of these specialized functions is now shown with an example.

## 10.3.1 `map_dbl()`

Here, we can process a number of columns from the data frame `mtcars` and return the average of each column as an atomic vector. Note that because a data frame is also a list, we can use it as an input to the `map` family of functions.

```
library(dplyr)
library(purrr)
mtcars %>%
  dplyr::select(mpg,cyl,disp) %>%
  purrr::map_dbl(mean)
#>      mpg      cyl     disp
#>   20.091    6.188  230.722
```

## 10.3.2 `map_chr()`

If we wanted to extract the director names from `sw_films` as an atomic vector, we can do this with `map_chr()`. Note that in this example, we use a formula to specify the function shortcut, and the argument is accessed using `.x`.

```
library(repurrrsive)
library(purrr)
sw_films %>%
  purrr::map_chr(~.x$director) %>%
  unique()
#> [1] "George Lucas"     "Richard Marquand" "Irvin Kershner"
#> [4] "J. J. Abrams"
```

## 10.3.3 `map_lgl()`

Here, we process a number of the columns in `mpg` to test whether the columns are numeric. Here, we use the anonymous function option.

```
library(ggplot2)
library(purrr)
```

```
library(dplyr)
mpg %>%
  dplyr::select(manufacturer:cyl) %>%
  purrr::map_lgl(function(x)is.numeric(x))
#> manufacturer          model         displ          year           cyl
#>        FALSE          FALSE          TRUE          TRUE          TRUE
```

### 10.3.4  `map_int()`

In this example, we select a number of numeric columns from `mpg`, and then use `map_int()` to count the number of observations that are greater than the mean in each of the three columns. An atomic vector of integers is returned.

```
library(ggplot2)
library(dplyr)
library(purrr)


mpg %>%
  dplyr::select(displ,cty,hwy) %>%
  purrr::map_int(~sum(.x>mean(.x)))
#> displ   cty   hwy
#>   107   118   129
```

### 10.3.5  `map_df()`

The function `map_df()` creates a new data frame or tibble based on the input list. A tibble is specified within the function, and as `map_df()` iterates through the input list, rows will be added to the tibble with the specified values. In this case, we generate a new tibble that extracts the Star Wars episode id, movie title, director, and release date. Note that we have converted the release date from a character vector to a date format.

```
library(repurrrsive)
library(purrr)
library(dplyr)


sw_films %>%
  purrr::map_df(~tibble(ID=.x$episode_id,
                        Title=.x$title,
                        Director=.x$director,
                        ReleaseDate=as.Date(.x$release_date))) %>%
  dplyr::arrange(ID)
#> # A tibble: 7 x 4
#>       ID Title                    Director          ReleaseDate
#>    <int> <chr>                    <chr>             <date>
```

```
#> 1    1 The Phantom Menace       George Lucas      1999-05-19
#> 2    2 Attack of the Clones     George Lucas      2002-05-16
#> 3    3 Revenge of the Sith      George Lucas      2005-05-19
#> 4    4 A New Hope               George Lucas      1977-05-25
#> 5    5 The Empire Strikes Back Irvin Kershner     1980-05-17
#> 6    6 Return of the Jedi      Richard Marquand 1983-05-25
#> 7    7 The Force Awakens        J. J. Abrams      2015-12-11
```

## 10.4    Iterating over two inputs using `map2()` and `pmap()`

The function map() iterates over a single argument .x, where that value could be
the columns in a data frame, or a set of elements in a list. However, there may
be cases where you need two input values as part of a data processing pipeline.
The function map2() allows for two inputs, and these are then represented as
arguments by .x and .y.

In this example, we can generate a stream of five normal random numbers with
three different options for the mean and standard deviation. These are stored
in the vectors means and sds. When passed to map2(), the elements of each
vector will be used together; for example, the first call will be rnorm(5,10,2),
and there will be three calls in all.

```
means <- c(10,20,30)
sds   <- c(2,4,7)

purrr::map2(means,sds,~rnorm(5,.x,.y)) %>% str()
#> List of 3
#>  $ : num [1:5] 10.42 6.78 10.54 12.05 7.1
#>  $ : num [1:5] 28 21.6 10.4 31.1 18.2
#>  $ : num [1:5] 32 35.3 30.6 25.5 28.3
```

But what if we need three inputs to a processing task? For example, the
function rnorm() requires three arguments, the number of random numbers
to be generated (n), the mean (mean), and the standard deviation (sd). While
there is no function called map3(), we can use another purrr function known as
pmap(), which can take a list containing any number of arguments, and process
these elements within the function using the symbols ..1, ..2 which represent
the first, second, and additional arguments.

```
params <- list(means = c(10,20,30),
               sds   = c(2,4,7),
               n     = c(4,5,6))
```

```
purrr::pmap(params,
            ~rnorm(n    = ..3,
                   mean = ..1,
                   sd   = ..2)) %>%
       str()
#> List of 3
#>  $ : num [1:4] 12.68 10.27 5.06 11.31
#>  $ : num [1:5] 20.1 18.5 21 14.9 20.2
#>  $ : num [1:6] 28 25 34.4 35.6 36.7 ...
```

We can also see how `pmap()` can be used to generate a summary of values for each row in a tibble.

Consider the variable `grades`, which shows synthetic data for six students, who receive marks in three different subjects. Note that, similar to the `map()` family of functions, `pmap()` also supports converting outputs to atomic vectors, so the function `pmap_chr()` is used.

```
set.seed(100)
grades <- tibble(ID=paste0("S-",10:15),
                 Subject1=rnorm(6,70,10),
                 Subject2=rnorm(6,60,20),
                 Subject3=rnorm(6,50,15))
grades
#> # A tibble: 6 x 4
#>    ID     Subject1 Subject2 Subject3
#>    <chr>    <dbl>    <dbl>    <dbl>
#> 1 S-10      65.0     48.4     47.0
#> 2 S-11      71.3     74.3     61.1
#> 3 S-12      69.2     43.5     51.9
#> 4 S-13      78.9     52.8     49.6
#> 5 S-14      71.2     61.8     44.2
#> 6 S-15      73.2     61.9     57.7
```

We now want to add a new column, using the `mutate()` function, that summarizes each student's grades in terms of the maximum grade received. In this case, we use the argument `..1` for the student ID, and the arguments `..2` to `..4` as input to the `max()` function.

```
grades1 <- grades %>%
           dplyr::mutate(Summary=pmap_chr(grades,
                              ~paste0("ID=",
                                      ..1,
                                      " Max=",
                                      round(max(..2,..3,..4),2))))
grades1
```

```
#> # A tibble: 6 x 5
#>    ID     Subject1 Subject2 Subject3 Summary
#>    <chr>  <dbl>    <dbl>    <dbl>    <chr>
#> 1 S-10     65.0     48.4     47.0   ID=S-10 Max=64.98
#> 2 S-11     71.3     74.3     61.1   ID=S-11 Max=74.29
#> 3 S-12     69.2     43.5     51.9   ID=S-12 Max=69.21
#> 4 S-13     78.9     52.8     49.6   ID=S-13 Max=78.87
#> 5 S-14     71.2     61.8     44.2   ID=S-14 Max=71.17
#> 6 S-15     73.2     61.9     57.7   ID=S-15 Max=73.19
```

A potential difficulty with this code is what might happen if the number of subject results changed; for example, if we had just two subjects instead of three.

To explore this scenario, we reduce the number of columns by setting Subject3 to NULL.

```
grades2 <- grades %>% dplyr::mutate(Subject3=NULL)
grades2
#> # A tibble: 6 x 3
#>    ID     Subject1 Subject2
#>    <chr>  <dbl>    <dbl>
#> 1 S-10     65.0     48.4
#> 2 S-11     71.3     74.3
#> 3 S-12     69.2     43.5
#> 4 S-13     78.9     52.8
#> 5 S-14     71.2     61.8
#> 6 S-15     73.2     61.9
```

Next, we modify the shortcut function within pmap(). Notice that we only explicitly reference the first argument ..1, and that the call list(...) can be used to get the complete set of arguments.

We create a list of grades by excluding the first argument from the list (the ID), and then flattening this to an atomic vector, before creating the desired string output.

```
grades2 <- grades2 %>%
            dplyr::mutate(Summary=pmap_chr(grades2,~{
                          args <- list(...)
                          grades <- unlist(args[-1])
                          paste0("ID=",..1," Max=",
                            round(max(grades),0))
                          }))
grades2
#> # A tibble: 6 x 4
#>    ID     Subject1 Subject2 Summary
```

```
#>     <chr>      <dbl>     <dbl> <chr>
#> 1 S-10        65.0      48.4 ID=S-10 Max=65
#> 2 S-11        71.3      74.3 ID=S-11 Max=74
#> 3 S-12        69.2      43.5 ID=S-12 Max=69
#> 4 S-13        78.9      52.8 ID=S-13 Max=79
#> 5 S-14        71.2      61.8 ID=S-14 Max=71
#> 6 S-15        73.2      61.9 ID=S-15 Max=73
```

## 10.5   Integrating `purrr` with `dplyr` and `tidyr` to process tibbles

A benefit of using the `tidyverse` is having the facility to combine tools from different packages, and switching between the use of lists and tibbles where appropriate. A common task is to divide a tibble into sub-groups, and perform operations on these. We have already seen how this can work using the package `dplyr`, which allows you to use the functions `group_by()` and `summarize()` to aggregate data.

As an illustration, we will explore two functions that can be used with `purrr` to process data, and we will focus on using these to perform statistical analysis, namely, correlations and linear models. We will explore the weather data in `aimsir17` to address the hypothesis of whether there is a strong association between atmospheric pressure and wind speed, and how this might vary across Ireland's weather stations.

The two functions we use with `purrr` are `dplyr::group_split()` and `tidyr::nest()`.

### 10.5.1   `group_split()`

This function, contained in the package `dplyr`, can be used to split a tibble into a *list of tibbles*, based on groupings specified by `group_by()`. This list can be processed using `map()`.

To show how this works, we first define a sample tibble from `mpg` that will have two different class values.

```
set.seed(100)
test <- mpg %>%
        dplyr::select(manufacturer:displ,cty,class) %>%
        dplyr::filter(class %in% c("compact","midsize")) %>%
        dplyr::sample_n(5)
test
```

```
#> # A tibble: 5 x 5
#>    manufacturer model  displ   cty class
#>    <chr>        <chr>  <dbl> <int> <chr>
#> 1 volkswagen   jetta   2      21 compact
#> 2 volkswagen   jetta   2.5    21 compact
#> 3 chevrolet    malibu  3.6    17 midsize
#> 4 volkswagen   gti     2      21 compact
#> 5 audi         a4      2      21 compact
```

Next, we can take this tibble, group it by class of car ("compact" and "midsize"), and then call group_split().

```
test_s <- test %>%
           dplyr::group_by(class) %>%
           dplyr::group_split()
test_s
#> <list_of<
#>   tbl_df<
#>     manufacturer: character
#>     model       : character
#>     displ       : double
#>     cty         : integer
#>     class       : character
#>   >
#> >[2]>
#> [[1]]
#> # A tibble: 4 x 5
#>    manufacturer model displ   cty class
#>    <chr>        <chr> <dbl> <int> <chr>
#> 1 volkswagen   jetta  2      21 compact
#> 2 volkswagen   jetta  2.5    21 compact
#> 3 volkswagen   gti    2      21 compact
#> 4 audi         a4     2      21 compact
#>
#> [[2]]
#> # A tibble: 1 x 5
#>    manufacturer model  displ   cty class
#>    <chr>        <chr>  <dbl> <int> <chr>
#> 1 chevrolet    malibu  3.6    17 midsize
```

The result is a list of tibbles, stored in the variable test_s. Note that each tibble contains all of the columns and rows for that particular class of vehicle.

We can now show how the output from group_split() integrates with map_int() to show the number of rows in each new tibble.

```
test_s %>% purrr::map_int(~nrow(.x))
#> [1] 4 1
```

A practical use of `group_split()` is now explored based on a weather data example from `aimsir17`, where we will use the following libraries.

```
library(aimsir17)
library(purrr)
library(dplyr)
```

Our goal is to calculate the correlation coefficient between two variables: mean sea level pressure and average wind speed. We simplify the dataset to daily values, where we take (1) the maximum wind speed (`wdsp`) recorded and (2) the average mean sea level pressure (`msl`). Our first task is to use `dplyr` to generate a summary tibble, and we also exclude any cases that have missing values, by combining `complete.cases()` within `filter`. Note that the function `complete.cases()` returns a logical vector indicating which rows are complete. The new tibble is stored in the variable `d_data`.

```
d_data <- observations %>%
           dplyr::filter(complete.cases(observations)) %>%
           dplyr::group_by(station,month,day) %>%
           dplyr::summarize(MaxWdsp=max(wdsp,na.rm=TRUE),
                            DailyAverageMSL=mean(msl,na.rm=TRUE)) %>%
           dplyr::ungroup()
d_data
#> # A tibble: 8,394 x 5
#>     station month    day MaxWdsp DailyAverageMSL
#>     <chr>   <dbl> <int>   <dbl>           <dbl>
#>  1 ATHENRY      1     1      12           1027.
#>  2 ATHENRY      1     2       8           1035.
#>  3 ATHENRY      1     3       6           1032.
#>  4 ATHENRY      1     4       4           1030.
#>  5 ATHENRY      1     5       9           1029.
#>  6 ATHENRY      1     6       9           1028.
#>  7 ATHENRY      1     7       6           1032.
#>  8 ATHENRY      1     8       9           1029.
#>  9 ATHENRY      1     9      16           1015.
#> 10 ATHENRY      1    10      13           1013.
#> # ... with 8,384 more rows
```

With this daily summary of data, we can now perform the correlation analysis. A key aspect of this is the use of the function `group_split()` which creates a list, where each list element contains a tibble for an individual station. Note the use of the `group_by()` call before we split the tibble.

```
cor7 <- d_data %>%
        dplyr::group_by(station) %>%
        dplyr::group_split() %>%
        purrr::map_df(~{
                corr <- cor(.x$MaxWdsp,.x$DailyAverageMSL)
                tibble(Station=first(.x$station),
                        CorrCoeff=corr)
        }) %>%
        dplyr::arrange(CorrCoeff) %>%
        dplyr::slice(1:7)
```

In this example, the function map_df() is used to format the output. A tibble is used, and has two columns, one for the weather station , and the second for the calculated correlation coefficient. The weather station is extracted using the function first(), which takes the first value in the tibble, given that all the entries for station within each group will be the same. The top seven are shown, and this indicates a moderate level of negative correlation between the variables for each weather station.

```
cor7
#> # A tibble: 7 x 2
#>    Station             CorrCoeff
#>    <chr>                   <dbl>
#> 1 SherkinIsland          -0.589
#> 2 VALENTIA OBSERVATORY   -0.579
#> 3 ROCHES POINT           -0.540
#> 4 MACE HEAD              -0.539
#> 5 MOORE PARK             -0.528
#> 6 SHANNON AIRPORT        -0.524
#> 7 CORK AIRPORT           -0.522
```

Note that in this particular example, we could also have used dplyr to calculate the result, and we can see what this code looks like. This is a familiar theme in R, where you may often find more than one way to perform the same task.

```
cor7_b <- d_data %>%
          dplyr::group_by(station) %>%
          dplyr::summarize(CorrCoeff=cor(MaxWdsp,DailyAverageMSL)) %>%
          dplyr::arrange(CorrCoeff) %>%
          dplyr::slice(1:7)
cor7_b
#> # A tibble: 7 x 2
#>    station             CorrCoeff
#>    <chr>                   <dbl>
#> 1 SherkinIsland          -0.589
#> 2 VALENTIA OBSERVATORY   -0.579
```

```
#> 3 ROCHES POINT            -0.540
#> 4 MACE HEAD               -0.539
#> 5 MOORE PARK              -0.528
#> 6 SHANNON AIRPORT         -0.524
#> 7 CORK AIRPORT            -0.522
```

In our next example, we will see another way for supporting a statistical workflow with purrr, using a different approach, but one that has some useful additional features.

### 10.5.2  nest()

The function nest(), which is part of the package tidyr, can be used to create a list column within a tibble that contains a data frame. Nesting generates one row for each defined group, which is identified using the function group_by(). The second column is named data, and is a list, and each list element contains all of the tibble's data for a particular group.

We can observe how nest() works taking a look at the weather example from the previous section.

```
data_n <- d_data %>%
            dplyr::group_by(station) %>%
            tidyr::nest()


data_n %>% head()
#> # A tibble: 6 x 2
#> # Groups:   station [6]
#>    station        data
#>    <chr>          <list>
#> 1 ATHENRY        <tibble [365 x 4]>
#> 2 BALLYHAISE     <tibble [365 x 4]>
#> 3 BELMULLET      <tibble [365 x 4]>
#> 4 CASEMENT       <tibble [365 x 4]>
#> 5 CLAREMORRIS    <tibble [365 x 4]>
#> 6 CORK AIRPORT   <tibble [365 x 4]>
```

Here, the tibble data_n contains two columns, with a row for each weather station (the first six rows are shown here). All of the data for each weather station is stored in the respective cell in the column data. For example, we can explore the data for "ATHENRY" with the following code. Note that the function first() is a wrapper around the list operator [[ that returns the first value in a list (the corresponding function last() returns the last list value).

```
data_n %>%
  dplyr::pull(data) %>%
  dplyr::first()
```

```
#> # A tibble: 365 x 4
#>    month   day MaxWdsp DailyAverageMSL
#>    <dbl> <int>   <dbl>           <dbl>
#>  1     1     1      12           1027.
#>  2     1     2       8           1035.
#>  3     1     3       6           1032.
#>  4     1     4       4           1030.
#>  5     1     5       9           1029.
#>  6     1     6       9           1028.
#>  7     1     7       6           1032.
#>  8     1     8       9           1029.
#>  9     1     9      16           1015.
#> 10     1    10      13           1013.
#> # ... with 355 more rows
```

Given that the new column created by nest() is a list, it can now be processed using map(), and, interestingly, we can use the mutate operation to store the results in a new column of the nested tibble. In particular, reverting to the weather example, we can run a linear regression model on each station's dataset, and store the results in a new column. In the linear regression model, the dependent variable is the maximum wind speed (MaxWdsp) and the independent variable is the average atmospheric pressure (DailyAverageMSL).

```
data_n <- data_n    %>%
            dplyr::mutate(LM=map(data,
                            ~lm(MaxWdsp~DailyAverageMSL,
                                data=.)))

data_n %>%
  head()
#> # A tibble: 6 x 3
#> # Groups:   station [6]
#>   station     data                LM
#>   <chr>       <list>              <list>
#> 1 ATHENRY     <tibble [365 x 4]> <lm>
#> 2 BALLYHAISE  <tibble [365 x 4]> <lm>
#> 3 BELMULLET   <tibble [365 x 4]> <lm>
#> 4 CASEMENT    <tibble [365 x 4]> <lm>
#> 5 CLAREMORRIS <tibble [365 x 4]> <lm>
#> 6 CORK AIRPORT <tibble [365 x 4]> <lm>
```

The tibble now contains the linear regression model result for each weather station, and therefore we have a complete set of model results. Given that the column LM is a list, we can extract any of the elements, for example, the summary of the model results for the weather station "BELMULLET".

```
data_n %>%
  dplyr::filter(station=="BELMULLET") %>%
  dplyr::pull(LM) %>%
  dplyr::first() %>%
  summary()
#>
#> Call:
#> lm(formula = MaxWdsp ~ DailyAverageMSL, data = .)
#>
#> Residuals:
#>     Min      1Q  Median      3Q     Max
#> -14.021  -4.069  -0.516   3.958  17.962
#>
#> Coefficients:
#>                 Estimate Std. Error t value Pr(>|t|)
#> (Intercept)      242.786     26.365    9.21   <2e-16 ***
#> DailyAverageMSL   -0.222      0.026   -8.53    4e-16 ***
#> ---
#> Signif. codes:  0 '***' 0.001 '**' 0.01 '*' 0.05 '.' 0.1 ' ' 1
#>
#> Residual standard error: 5.8 on 363 degrees of freedom
#> Multiple R-squared:  0.167,  Adjusted R-squared:  0.165
#> F-statistic: 72.8 on 1 and 363 DF,  p-value: 4.03e-16
```

To continue our exploration of dplyr and purrr, we can now add a new column RSq' that contains the $R^2$ value for each model.

```
data_n <- data_n %>%
            dplyr::mutate(RSq=map_dbl(LM,~summary(.x)$r.squared)) %>%
            dplyr::arrange(desc(RSq))
data_n   <- data_n %>% head(n=3)
data_n
#> # A tibble: 3 x 4
#> # Groups:    station [3]
#>   station              data                LM       RSq
#>   <chr>                <list>              <list>   <dbl>
#> 1 SherkinIsland        <tibble [365 x 4]>  <lm>     0.347
#> 2 VALENTIA OBSERVATORY <tibble [365 x 4]>  <lm>     0.335
#> 3 ROCHES POINT         <tibble [365 x 4]>  <lm>     0.291
```

This shows the three stations with the highest r-squared values. While an in-depth discussion of the significance of these is outside the scope of this book, the key point is that we have generated statistical measures via a data processing pipeline that is made possible by using the tools of purrr, dplyr, and tidyr.

We can now do one final activity on the data by revisiting the original data for these three stations, and then plotting this on a graph, shown in Figure 10.2. This code makes use of the geom_smooth() function to show the linear models.

```
data1 <- d_data %>%
        dplyr::filter(station %in% dplyr::pull(data_n,station))

ggplot(data1,aes(x=DailyAverageMSL,y=MaxWdsp,color=station))+
        geom_point()+geom_smooth(method="lm")+geom_jitter()
```

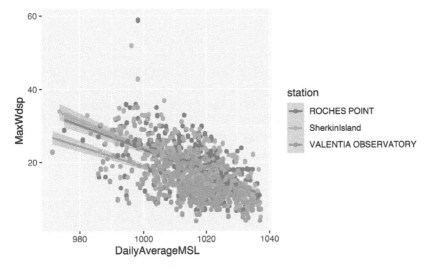

**FIGURE 10.2** Exporing relationships between atmospheric pressure and wind speed

## 10.6 Additional **purrr** functions

We now summarize additional useful functions provided by purrr.

### 10.6.1 **pluck()**

The function pluck() provides a generalized form of the [[ operator and provides the means to index data structures in a flexible way. The arguments include .x, which is a vector, and a list of accessors for indexing into the object, which can include an integer position or a string name. Here are some examples.

```r
library(ggplot2)
library(repurrrsive)

# Use pluck() to access the second element of an atomic vector
mpg %>% dplyr::pull(class) %>% unique() %>% purrr::pluck(2)
#> [1] "midsize"
# Use pluck() to access the director in the first list location
sw_films %>% purrr::pluck(1,"director")
#> [1] "George Lucas"
```

### 10.6.2  walk()

The function walk(.x,.f) is similar to map, except that it returns the input .x
and calls the function .f to generate a side effect. The side effect, for example,
could be displaying information onto the screen, and no output value needs to
be returned. Here is a short example.

```r
l <- list(el1=20,el2=30,el3=40)
o <- purrr::walk(l,~cat("Creating a side effect...\n"))
#> Creating a side effect...
#> Creating a side effect...
#> Creating a side effect...
str(o)
#> List of 3
#>  $ el1: num 20
#>  $ el2: num 30
#>  $ el3: num 40
```

### 10.6.3  keep()

The function keep(.x,.f) takes in a list .x and, based on the evaluation of a
predicate function, will either keep or discard list element. In effect, it provides
a way to filter a list. Here, we can filter those movies that have George Lucas
as a director, and then confirm the result using walk().

```r
o <- sw_films %>% keep(~.x$director=="George Lucas")
purrr::walk(o,~cat(.x$director," ==> Title =",.x$title,"\n"))
#> George Lucas  ==> Title = A New Hope
#> George Lucas  ==> Title = Attack of the Clones
#> George Lucas  ==> Title = The Phantom Menace
#> George Lucas  ==> Title = Revenge of the Sith
```

### 10.6.4 `invoke_map()`

The function `invoke_map()` provides a feature to call a list of functions with a list of parameters. It is a wrapper around the R function `do.call()`. The function takes a list of functions (where the function names are contained in a character vector), and an argument list, where the names of the function arguments are contained in the list. Here, we can see an example of where a vector is passed into a number of different functions.

```
f <- c("min","max","sum")
l <- list(
  arg1=list(x=1:3),
  arg2=list(x=10:12),
  arg3=list(x=1:4)
)
```

```
str(purrr::invoke_map(f,l))
#> Warning: `invoke_map()` was deprecated in purrr 1.0.0.
#> i Please use map() + exec() instead.
#> This warning is displayed once every 8 hours.
#> Call `lifecycle::last_lifecycle_warnings()` to see where this
#> warning was generated.
#> List of 3
#>  $ : int 1
#>  $ : int 12
#>  $ : int 10
```

## 10.7 Mini-case: Generating linear models from the mpg dataset

The aim of this mini-case is to illustrate how the tools of `purrr`, `dplyr`, and `tidyr` can be used to support a statistical modelling workflow, and to store the results in a convenient way. The idea is to generate a collection of linear models from the `mpg` tibble, and explore the relationship between a dependent variable (`cty`) and a dependent variable (`displ`). In this example, we will partition the data by vehicle class, and so build seven different linear models. Before starting the modelling workflow, we can visualize, in Figure 10.3, the different linear models using `ggplot()`. We also include the required libraries for our analysis.

```
library(purrr)
library(tidyr)
library(ggplot2)
ggplot(mpg,aes(x=displ,y=cty))+
```

```
geom_point()+geom_smooth(method="lm")+
facet_wrap(~class)
```

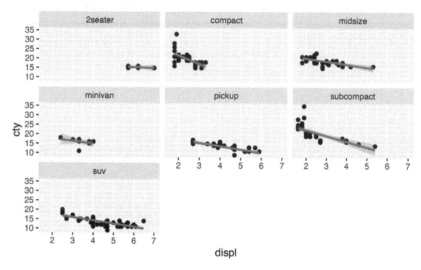

**FIGURE 10.3** The relationship between displacement and city miles per gallon

Our overall approach is as follows:

- Create a nested tibble (grouped by class), and this will have two columns, `class` and `data`, where the `data` column is created by the call to `nest()`.
- Using `mutate()`, add a new column that will store the results of each linear model.
- Process the linear model results to extract a measure of model performance (r-squared) to see how well each regression model explains observed data.
- Arrange the tibble so that the best model scores are shown first.
- Add a column that will store the plots for each model.

First, we transform `mpg` using `nest()`. Note that the `nest()` function generates a new tibble with one row for each class of variable, and the remaining data for that variable is stored in a new column as a list. This is a valuable tibble property, as it can store a complex data structure such as a list in a cell value. To reverse the "nesting process", simply call the related function `unnest()`.

```
mpg1 <- mpg %>%
        dplyr::group_by(class) %>%
        tidyr::nest()
mpg1
#> # A tibble: 7 x 2
#> # Groups:   class [7]
#>   class        data
```

```
#>    <chr>        <list>
#> 1 compact      <tibble [47 x 10]>
#> 2 midsize      <tibble [41 x 10]>
#> 3 suv          <tibble [62 x 10]>
#> 4 2seater      <tibble [5 x 10]>
#> 5 minivan      <tibble [11 x 10]>
#> 6 pickup       <tibble [33 x 10]>
#> 7 subcompact   <tibble [35 x 10]>
```

The data for each vehicle class can be extracted, and using the pull() (similar to $) function followed by first(). We can view the first tibble, which is for the class "compact".

```
mpg1 %>%
  dplyr::pull(data) %>%
  dplyr::first()
#> # A tibble: 47 x 10
#>    manufa~1 model displ  year   cyl trans drv     cty   hwy fl
#>    <chr>    <chr> <dbl> <int> <int> <chr> <chr> <int> <int> <chr>
#>  1 audi     a4      1.8  1999     4 auto~ f        18    29 p
#>  2 audi     a4      1.8  1999     4 manu~ f        21    29 p
#>  3 audi     a4      2    2008     4 manu~ f        20    31 p
#>  4 audi     a4      2    2008     4 auto~ f        21    30 p
#>  5 audi     a4      2.8  1999     6 auto~ f        16    26 p
#>  6 audi     a4      2.8  1999     6 manu~ f        18    26 p
#>  7 audi     a4      3.1  2008     6 auto~ f        18    27 p
#>  8 audi     a4 q~   1.8  1999     4 manu~ 4        18    26 p
#>  9 audi     a4 q~   1.8  1999     4 auto~ 4        16    25 p
#> 10 audi     a4 q~   2    2008     4 manu~ 4        20    28 p
#> # ... with 37 more rows, and abbreviated variable name
#> #   1: manufacturer
```

Using mutate() and map(), we add a new column to mpg1 that will store the results of each linear regression model.

```
mpg1 <- mpg1 %>%
        dplyr::mutate(LM=map(data,~lm(cty~displ,data=.x)))
mpg1
#> # A tibble: 7 x 3
#> # Groups:   class [7]
#>   class    data                 LM
#>   <chr>    <list>               <list>
#> 1 compact  <tibble [47 x 10]> <lm>
#> 2 midsize  <tibble [41 x 10]> <lm>
#> 3 suv      <tibble [62 x 10]> <lm>
#> 4 2seater  <tibble [5 x 10]>  <lm>
```

```
#> 5 minivan     <tibble [11 x 10]> <lm>
#> 6 pickup      <tibble [33 x 10]> <lm>
#> 7 subcompact <tibble [35 x 10]> <lm>
```

The map function operates on the `data` column of `mpg1`, which is a list, and therefore is a suitable input for `map()`. The shorter notation for a function is used, and the `.x` value will represent the tibble for each class of car. The independent variable `displ` is selected, along with the dependent variable `cty`, and the function `lm` is called, which returns an "lm" object. This object is then returned for every iteration through the list and is stored in a new column `LM`.

Therefore, the tibble `mpg1` now has the full set of modelling results for each class of car, and this can then be used for additional processing.

We can explore the linear models by operating on the tibble; for example, the following command shows a summary of the model for class "suv".

```
mpg1 %>%
   dplyr::filter(class=="suv") %>%    # Select row where class == suv"
   dplyr::pull(LM) %>%                # Get the column "LM"
   dplyr::first() %>%                 # Extract the lm object
   summary()                          # Call the summary function
#>
#> Call:
#> lm(formula = cty ~ displ, data = .x)
#>
#> Residuals:
#>     Min     1Q Median    3Q    Max
#> -4.087 -1.027 -0.087 1.096  3.967
#>
#> Coefficients:
#>              Estimate Std. Error t value Pr(>|t|)
#> (Intercept)    21.060      0.893    23.6  < 2e-16 ***
#> displ          -1.696      0.195    -8.7 3.2e-12 ***
#> ---
#> Signif. codes:  0 '***' 0.001 '**' 0.01 '*' 0.05 '.' 0.1 ' ' 1
#>
#> Residual standard error: 1.62 on 60 degrees of freedom
#> Multiple R-squared:  0.558,  Adjusted R-squared:  0.55
#> F-statistic: 75.7 on 1 and 60 DF,  p-value: 3.17e-12
```

The `summary()` function also returns a list containing information relating to the linear regression process. This list includes the element "r.squared", which is a measure of how well each regression model explains observed data. We can now add this information to our tibble `mpg1`, and re-arrange the tibble from highest to lowest r-squared values, this providing comparative information on the seven models.

```
mpg1 <- mpg1 %>%
        dplyr::mutate(RSquared=map_dbl(LM,~summary(.x)$r.squared)) %>%
        dplyr::arrange(desc(RSquared))
mpg1
#> # A tibble: 7 x 4
#> # Groups:   class [7]
#>   class      data                 LM     RSquared
#>   <chr>      <list>               <list>   <dbl>
#> 1 suv        <tibble [62 x 10]> <lm>       0.558
#> 2 subcompact <tibble [35 x 10]> <lm>       0.527
#> 3 pickup     <tibble [33 x 10]> <lm>       0.525
#> 4 compact    <tibble [47 x 10]> <lm>       0.358
#> 5 midsize    <tibble [41 x 10]> <lm>       0.339
#> 6 2seater    <tibble [5 x 10]>  <lm>       0.130
#> 7 minivan    <tibble [11 x 10]> <lm>       0.124
```

To conclude our look at a data modelling workflow, we can also store custom plots in the output tibble, so that they can be extracted at a later stage. To visualize the plots we will use the R package `ggpubr`, which contains a valuable function called `ggarrange()` that allows you to plot a list of ggplot objects. Before we complete this task, we reiterate the point that a ggplot can be stored in a variable and is actually a list which belongs to the S3 class "gg", which inherits from the S3 class "ggplot".

```
p <- ggplot(mpg,aes(x=displ,y=cty))+geom_point()
typeof(p)
#> [1] "list"
names(p)
#> [1] "data"       "layers"       "scales"   "mapping"
#> [5] "theme"      "coordinates" "facet"     "plot_env"
#> [9] "labels"
class(p)
#> [1] "gg"       "ggplot"
```

This means that we can store all the plots in a tibble column and create this column using a combination of `mutate()` and `map2()`. Note that the column `Plots` in `mpg1` will contain a list of the seven ggplots.

```
library(randomcoloR) # Functions to generate different colors
set.seed(100) # For random color generation

mpg1 <- mpg1 %>%
        dplyr::mutate(Plots=map2(class,data, ~{
                # (1) run linear model
                m <- lm(cty~displ,data=.y)
                # (2) Extract coefficients
```

```
intercept <- round(coef(m)[1],2)
slope     <- round(coef(m)[2],2)
# (3) return the ggplot,
#       include the #obs and
#       coefficients in title
ggplot(.y,aes(x=displ,y=cty))+
  ggtitle(paste0(.x,"(#",nrow(.y),") I=",
                      intercept,
                      " S=",slope))+
  geom_point(color=randomColor(),size=1.5)+
  geom_abline(slope=slope,intercept = intercept)+
  theme_classic()+
  geom_jitter()+
  theme(plot.title = element_text(size = 7,
                          face = "italic"))
}))
```

The new column on the mpg1 tibble can now be viewed, and this column contains all the information needed to plot each graph.

```
mpg1
#> # A tibble: 7 x 5
#> # Groups:   class [7]
#>   class       data              LM     RSquared Plots
#>   <chr>       <list>            <list>    <dbl> <list>
#> 1 suv         <tibble [62 x 10]> <lm>     0.558 <gg>
#> 2 subcompact  <tibble [35 x 10]> <lm>     0.527 <gg>
#> 3 pickup      <tibble [33 x 10]> <lm>     0.525 <gg>
#> 4 compact     <tibble [47 x 10]> <lm>     0.358 <gg>
#> 5 midsize     <tibble [41 x 10]> <lm>     0.339 <gg>
#> 6 2seater     <tibble [5 x 10]>  <lm>     0.130 <gg>
#> 7 minivan     <tibble [11 x 10]> <lm>     0.124 <gg>
```

To display the plots, we use the function ggarrange(). This can take a list of plots as an input (plotlist), and the column Plots is passed through as the data for this argument. Figure 10.4 displays the purpose-built plot that encompasses: (1) the class of car, the number of observations, and the regression coefficients, and (2) the line of best fit, which was created using the ggplot2 function geom_abline().

```
p1 <- ggarrange(plotlist=pull(mpg1,Plots))
p1
```

In summary, this example shows the utility of combining the tools of purrr, dplyr, and tidyr in a workflow, specifically:

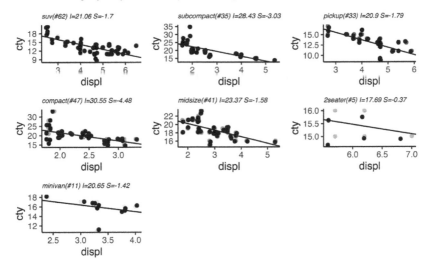

**FIGURE 10.4** Creating our own plots to summarize the relationships

- From the original tibble mpg, a new nested tibble (mpg1) was created that stored the data for each class in a single cell. The nest() function collapses the tibble into one row per group identified.
- Combining mutate() and map(), a new column was created that contained the result of a linear model where displ was the independent variable and cty the dependent variable. This result was essentially the "lm" object returned by the lm() function.
- The LM object was then used to create a new column showing the $R^2$ value for each model.
- Finally, the original two columns in mpg1 were used to generate special purpose ggplot graphs for each regression task, with informative titles added to each plot.

## 10.8   Summary of R functions from Chapter 10

The most important relational functions from the package purrr are now summarized, along with other functions called in the chapter.

| Function | Description |
| --- | --- |
| complete.cases() | Identifies cases with no missing values. |
| map() | Applies a function to each input, returns a list. |
| map_dbl() | Applies a function to each input, returns a double. |
| map_chr() | Applies a function to each input, returns a character. |
| map_lgl() | Applies a function to each input, returns a logical. |
| map_int() | Applies a function to each input, returns an integer. |
| map_df() | A map operation that returns a tibble. |
| map2() | A map operation that takes two input arguments. |
| pmap() | A map operation that takes many input arguments. |
| group_split() | Splits a data frame into a list of data frames. |
| nest() | Creates a list column of tibbles, based on group_by(). |
| pluck() | Implements a form of the [[ operator for vectors. |
| walk() | Returns the input, enables side effects. |
| keep() | Keeps elements that are evaluated as TRUE. |
| ggarrange() | Arranges multiple ggplots in the same page (ggpubr). |

## 10.9 Exercises

1. Create the following tibble with the three columns shown, using the functions `keep()` and `map_df()`, in order to provide a tabular view of the list `repurrrsive::sw_vehicles`. Note that possible invalid values of `length` in `sw_vehicles` include "unknown", and any of these should be removed prior to creating the data frame.

```
#> # A tibble: 6 x 3
#>   Name                   Model                            Length
#>   <chr>                  <chr>                            <dbl>
#> 1 C-9979 landing craft   C-9979 landing craft             210
#> 2 SPHA                   Self-Propelled Heavy Artillery   140
#> 3 Clone turbo tank       HAVw A6 Juggernaut               49.4
#> 4 Sand Crawler           Digger Crawler                   36.8
#> 5 Multi-Troop Transport  Multi-Troop Transport            31
#> 6 Sail barge             Modified Luxury Sail Barge        30
```

2. Explore how the function `invoke_map()` can be used to take the following input, which includes a list of parameters, and a vector of function names.

```
set.seed(1000)
params <- list(
```

```
  list(n=3,mean=10,sd=2),
  list(n=4,min=10,max=15),
  list(n=5,min=0,max=1)
)
f <- c("rnorm","runif","runif")
```

The following output should be generated.

```
str(res)
#> List of 3
#>  $ : num [1:3] 9.11 7.59 10.08
#>  $ : num [1:4] 13.7 12.9 11.1 11.3
#>  $ : num [1:5] 0.35 0.755 0.317 0.866 0.764
```

3. Generate the following daily summaries of rainfall and mean sea level pressure, for all the weather stations in `aimsir17::observations`, and only consider observations with no missing values.

```
#> `summarize()` has grouped output by 'station', 'month'. You can
#> override using the `.groups` argument.
d_sum
#> # A tibble: 8,394 x 5
#>     station month   day TotalRain AvrMSL
#>     <chr>   <dbl> <int>     <dbl>  <dbl>
#>  1 ATHENRY      1     1       0.2  1027.
#>  2 ATHENRY      1     2       0    1035.
#>  3 ATHENRY      1     3       0    1032.
#>  4 ATHENRY      1     4       0    1030.
#>  5 ATHENRY      1     5       0.1  1029.
#>  6 ATHENRY      1     6      18    1028.
#>  7 ATHENRY      1     7       1.4  1032.
#>  8 ATHENRY      1     8       1.2  1029.
#>  9 ATHENRY      1     9       5.4  1015.
#> 10 ATHENRY      1    10       0.7  1013.
#> # ... with 8,384 more rows
```

Next, using the tibble `d_sum` as input, generate the top 6 correlations between `TotalRain` and `AvrMSL` using `group_split()` and `map_df()`. Here are the results you should find.

```
cors
#> # A tibble: 6 x 2
#>    Station           Corr
#>    <chr>            <dbl>
#> 1 MOORE PARK      -0.496
#> 2 MULLINGAR       -0.496
```

```
#> 3 CLAREMORRIS              -0.484
#> 4 VALENTIA OBSERVATORY -0.464
#> 5 KNOCK AIRPORT            -0.459
#> 6 OAK PARK                 -0.456
```

4. Re-implement the following function as `my_map_dbl1()` and make use of the function `pluck()` instead of `f(x[[i]])`.

```
my_map_dbl1 <- function(x,f){
  out <- vector(mode="list",length=length(x))
  for(i in seq_along(x)){
    out[[i]] <- f(x[[i]])
  }
  unlist(out)
}
```

```
my_map_dbl1(1:3,function(x)x+10)
#> [1] 11 12 13
```

```
#> [1] 11 12 13
```

Here is the sample output from `my_map_dbl2()` which, as expected, is the same as the output from `my_map_dbl1()`.

```
my_map_dbl2(1:3,function(x)x+10)
#> [1] 11 12 13
```

5. The aim of this exercise is to analyze electricity demand data from Ireland in 2017, via the `eirgrid17` tibble. The first task is to generate the following tibble `e17`, based on the average daily demand using the column `IEDemand`. The `Day` column should be added through the function `lubridate::wday()`, with the arguments `label` and `abbr` both set to `TRUE`.

```
e17
#> # A tibble: 365 x 5
#>    month   day Date                Day   AvrDemand
#>    <dbl> <int> <dttm>              <ord>     <dbl>
#> 1      1     1 2017-01-01 00:00:00 Sun       2818.
#> 2      1     2 2017-01-02 00:00:00 Mon       3026.
#> 3      1     3 2017-01-03 00:00:00 Tue       3516.
#> 4      1     4 2017-01-04 00:00:00 Wed       3579.
#> 5      1     5 2017-01-05 00:00:00 Thu       3563.
#> 6      1     6 2017-01-06 00:00:00 Fri       3462.
#> 7      1     7 2017-01-07 00:00:00 Sat       3155.
#> 8      1     8 2017-01-08 00:00:00 Sun       2986.
```

```
#>  9      1      9 2017-01-09 00:00:00 Mon        3423.
#> 10      1     10 2017-01-10 00:00:00 Tue        3499.
#> # ... with 355 more rows
```

Next, use the functions `tidyr::nest()` and `purrr::map_dbl()` to create the following summary of energy demand, by day of the week.

```
e17_n
#> # A tibble: 7 x 3
#> # Groups:    Day [7]
#>   Day    data                AvrDemand
#>   <ord>  <list>                  <dbl>
#> 1 Thu    <tibble [52 x 4]>       3298.
#> 2 Wed    <tibble [52 x 4]>       3293.
#> 3 Tue    <tibble [52 x 4]>       3270.
#> 4 Fri    <tibble [52 x 4]>       3259.
#> 5 Mon    <tibble [52 x 4]>       3182.
#> 6 Sat    <tibble [52 x 4]>       3007.
#> 7 Sun    <tibble [53 x 4]>       2866.
```

# 11

## *Shiny*

Spreadsheets are closely related to reactive programming: you declare the relationship between cells using formulas, and when one cell changes, all of its dependencies automatically update.

— Hadley Wickham (Wickham, 2021)

## 11.1 Introduction

The Shiny system in R provides an innovative framework where R code can be used to create interactive web pages. These pages can take inputs from users, and then process these to generate a range of dynamic outputs, including text summaries, tables, and visualizations. The web page components are generated from standard R outputs, such as tibbles and plots. The R package shiny enables you to easily define a web page with inputs and outputs, and then write a special server function that can *react* to changes in inputs, and render these changes to web page outputs. While Shiny is a highly detailed topic in itself, in this chapter the emphasis is to provide an initial overview through a set of six examples which illustrate how Shiny works. Upon completion of this chapter, you should understand:

- The main concept behind reactive programming, the parallels with spreadsheets and how Shiny generates interactive content.
- The three aspects of a Shiny program: the ui variable that defines the interface using fluidPage(), the server() function that controls the behavior of the page, and the call to shinyApp() which creates the Shiny app objects.
- How to use the input Shiny controls: numericInput(), selectInput(), checkboxInput().

- How to access input controls within the `server()` function, and how to provide information for output sections of the web page, using the `render*()` family of functions.
- How to send data to `textOutput()` using the `renderText()` function.
- How to send data to `verbatimTextOutput()` using the `renderPrint()` function.
- How to send tibble information to `tableOutput()` using the `renderTable()` function.
- How to send graphical information to `plotOutput()` using the `renderPlot()` function.

**Chapter structure**

- Section 11.2 introduces the idea of reactive programming, using the example of a spreadsheet.
- Section 11.3 presents a simple Shiny example, which prints a welcome message to a web page.
- Section 11.4 focuses on squaring an input number.
- Section 11.5 demonstrates how to return a tibble summary based on a selected input value.
- Section 11.6 allows the user to compare summary results from two weather stations from the package `aimsir17`.
- Section 11.7 shows how a scatterplot can be configured from a selection of variables.
- Section 11.8 presents the idea of a *reactive expression*, and how these can be used to improve code efficiency.
- Section 11.9 provides a summary of all the functions introduced in the chapter.
- Section 11.10 provides short coding exercises to test your understanding of Shiny.

---

## 11.2 Reactive programming

By the end of the chapter, you should have an appreciation of how Shiny can be viewed as a reactive programming language, where relationships are declared between components, and when a component (i.e., an input) changes, all outputs that are dependent on the input will be re-calculated. An example of this process is a spreadsheet, where the user can establish formula dependencies between cells. We show this process in Figure 11.1, where we have two cells, A1 and B1, and B1's value depends on A1 (i.e. B1 = A1*2).

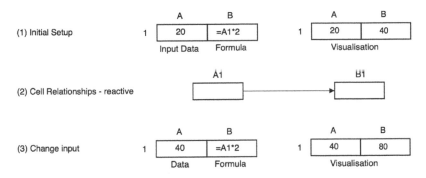

**FIGURE 11.1** A spreadsheet as an example of reactive programming

We can summarize the spreadsheet's reactive programming process as follows:

1. We program the spreadsheet's logic, where A1 contains the value 20, and B1 holds a formula, which is twice the value of A1. Therefore, we see the values of 20 for A1 and 40 for B1.

2. We view the implicit relationship between the cells. Later on in the chapter, we apply this idea to Shiny and will call it a *reactive graph*. Here, we simply draw a line pointing from A1 to B1, and this represents the scenario that any change in A1 will, in turn, lead to a change in B1. The spreadsheet internal design ensures that this happens.

3. We show that a change in A1 then triggers a change in B1's displayed value. We change A1 to 40, and the spreadsheet then reacts to change B1 to 80.

So while this is a simple example, it's also instructive, as when we create our web pages within Shiny, a similar mechanism will be used by Shiny to link web page inputs to outputs. This structure will facilitate the design of interactive and dynamic web apps. We now explore six examples, starting with the simplest case.

## 11.3 Example one: Hello Shiny

Inspired by many classic programmming textbooks, we start with an example that prints the text, "Hello, Shiny!" to a web page. This is a *static web page*, in that it does not respond to any user input, and it displays the same content every time it is loaded. We first load the shiny library, which contains all the functions that allows us to specify a basic web page, without having to know any web programming or scripting languages, such as HTML, CSS, or JavaScript.

```
library(shiny)
```

With Shiny, normally two objects are declared. One for the appearance of the app, which here is named ui, and here we call the function fluidPage() to create the web page, and provide a layout. A fluid page essentially contains a sequence of rows, and this example has one row, which returns the string "Hello, Shiny!". The function message() is used to display messages in the developer's console, and it ensures that, as developers, we can see detailed messages of what functions are being called.

```
ui <- fluidPage(
  message("Starting the UI..."),
  "Hello, Shiny!"
)
```

The second object required for a Shiny app is a user-defined function, usually named server, which takes in three parameters, two of which we use in the remaining examples.

- input, which provides direct access to an input control object defined in the function fluidPage().
- output, which provides access to an output component that has been defined in the function fluidPage().

Therefore, the server function *makes things happen* and allows you to write code to react to user input and generate output. In essence, the server function allows you to make an app interactive. Our first server is simple enough, and no actions are taken to read inputs or send back outputs.

```
server <- function(input, output, session){
  message("\nStarting the server...\n")
}
```

The final task is to call the function shinyApp() which takes in the user interface and server objects, and creates the web page. It performs work behind the scenes to make this happen. This is a benefit of Shiny, as you can focus on defining inputs and outputs, and then implement actions in the server function to ensure interactivity.

```
shinyApp(ui, server)
```

By clicking on the button "Run App" on the RStudio interface, the app will be launched, and something similar to the the output in Figure 11.2 should appear.

**FIGURE 11.2** The web page produced for the Hello Shiny code

There are a couple of points worth noting:

- The output will indicate the port number that the web server is running on, in this case it is port 6908 on a local machine.

- We can see the text output that we specified - "Hello, Shiny" - appearing on the first output line of the web page.

We will now go on to add a basic level of interaction to our next Shiny example.

## 11.4 Example two: Squaring an input number

This example involves taking input from the user (a number), squaring it, and then displaying the result. To start with, we access a new library as well as the library shiny. The library glue is used to concatenate strings, and configure meaningful messages that are then relayed to the web page.

```
library(shiny)
library(glue)
```

We then create our user interface object with a call to fluidPage(), and within this function we call two Shiny functions to capture input and display output:

- numericInput(), which will take in a number from the user, where the default value is 10. We add a label for the numeric input, which in this case is called "input_num". This will be used by the server code, so that the server can then access the value entered by the user. The message displayed to the user is also set ("Enter Number").

- textOutput(), which will be used to render a reactive output variable within the application page. This is given a label, "msg", so that when we write the server code we can re-direct output to this element of the web page by using the value "msg".

So in effect, this code specifies two rows for the web page. The first row numericInput() will allow the user to provide an input. The second row textOutput() will render the server function's output variable as text within an application page.

```
ui <- fluidPage(
  numericInput("input_num",
               "Enter Number",
               10),
  textOutput("msg")
)
```

Next, we implement our first server function. There are two aspects to consider:

- First, the argument `output` is used to access an output component in the web page, and in this case, it is `output$msg`, which references the `textOutput()` part of the web page that was defined in the function call to `fluidPage()`.

- Second, for each output function defined in the user interface, there is a corresponding function that needs to be defined in the server, and this function will render the information to its correct format. For the function `textOutput()`, the appropriate render function used is `renderText()`, and, just as with any function, the evaluated expression is returned. In this case, the function `glue()` is used to assemble the final output, which is in string format.

```
server <- function(input, output, session){
  output$msg<- renderText({
    message(glue("Input is {input$input_num}"))
    ans <- input$input_num^2
    glue("The square of {input$input_num} is {ans}\n")
  })
}
```

Therefore, stepping through the code within `server()` we can see that the variable `output$msg` is assigned its value within the function call to `renderText()`. Inside this function, there is one main processing step, where the variable `input$input_num` is squared, and the result stored in `ans`. This variable `ans` is then used to configure the output string through the function `glue()`, where curly braces are used to identify variables.

With the code completed, we define the two components of the app, and run the code.

```
shinyApp(ui, server)
```

The resulting web page is shown in Figure 11.3, and it shows the numeric input component (with a default value of 10), and then the output, which is configured by the call to the function `glue()` in the server.

Interestingly, in the figure, we also present a *reactive graph*, which echoes the spreadsheet example from earlier and shows the relationship between the input and output. Essentially, any time an input changes, the output will then change

(a) Shiny web page and components

http://127.0.0.1:6908   🔊 **Open in Browser**   ⟳

**Enter Number**

| 10 |
| :-- |

The square of 10 is 100

(b) Reactive graph showing how the input and output are connected

**FIGURE 11.3** A simple app that squares an input number

(via the server code). Shiny ensures that all the correct work is accomplished behind the scenes to make this reactive programming process possible.

## 11.5   Example three: Exploring a weather station

Our third example summarizes data from any one of six weather stations, based on the tibble `observations` from the package `aimsir17`. We first define the libraries that will be used for the web page.

```
library(shiny)
library(aimsir17)
library(dplyr)
library(glue)
```

Next, we create the list of six stations so that these can be added in the new input object, which is a selection list. The data processing pipeline is shown below, and the function `pull()` is used to convert a one-column tibble to a vector. The unique values are returned, which would include all 25 stations, and the function `head()` will then select the first six.

```
stations6 <- observations %>%
              dplyr::select(station) %>%
              dplyr::pull() %>%
              unique() %>%
              head()
stations6
#> [1] "ATHENRY"      "BALLYHAISE"   "BELMULLET"    "CASEMENT"
#> [5] "CLAREMORRIS"  "CORK AIRPORT"
```

In constructing the user interface, we have three main components:

- A title panel, which allows you to give the web page a title.
- The `selectInput()` function, whose label is "station", for creating a list that allows the user to select one of the six weather stations.
- A text output component, `verbatimTextOutput()`, which renders the output variable as text within the page. In this example, it will be the output from the `summary()` function. The label for the output is "summary".

```
ui <- fluidPage(
    message("\nStarting the UI..."),
    titlePanel("Exploring Weather Stations"),
    selectInput("station",
                label="Weather Stations",
                choices=stations6),
    verbatimTextOutput("summary")
)
```

The server function can now be written. In this case, as we are only targeting one output, and this output is named "summary", we allocate the result to the variable `output$summary`. The appropriate render function for `verbatim-TextOutput()` is `renderPrint()`, and inside of the render function call we add the following R code:

- Using `dplyr::filter()`, we filter the observations for the station selected by the user, and we access this value using the variable `input$station`.

- Six columns are selected, just so that they will fit on one line of output.

- The R function `summary()` is called, and its value returned as output.

```
server <- function(input, output, session){

    message("\nStarting the server...")

    output$summary <- renderPrint({
        message(glue("output$summary>> input$station = {input$station}"))
        dataset <- dplyr::filter(observations,
                                 station == input$station) %>%
                   dplyr::select(rain,
                                 temp,
                                 rhum,
                                 msl,
                                 wdsp,
                                 wddir)
        cat("Summary Output for ",input$station,"\n\n")
        summary(dataset)
    })

}
```

We then call the `shinyApp()` function to specify the user interface and server objects.

**shinyApp**(ui, server)

The app is then run, and the web page is shown in Figure 11.4. It shows the title panel, which is followed by the selection input control. The default in the selection list is the first item on the list of six. Finally, the output is shown, which is generated inside the `server()` function. It produces output as would be shown in the R console, and this shows the benefit of the `verbatimTextOutput()` component for displaying R output.

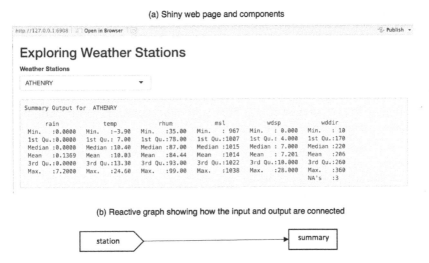

(a) Shiny web page and components

(b) Reactive graph showing how the input and output are connected

**FIGURE 11.4** Summarizing weather station variables from aimsir17

We show the reactive graph, which captures the relationship between the input variable component ("station") and the output component ("summary"). This diagram is informative, as it shows that our program will react to a change in "station", by changing the output "summary".

## 11.6 Example four: Comparing two weather stations

The fourth example involves comparing the monthly rainfall values for two weather stations. The six stations from the previous example are made available for selection, and two list selection input components are used. As a first step, the required libraries are loaded, `tidyr` is included as we will need to widen data in order to present it in a user-friendly manner.

```
library(shiny)
library(aimsir17)
library(dplyr)
library(ggplot2)
library(glue)
library(tidyr)
```

Similar to the previous example, we prepare six weather stations for selection.

```
stations6 <- observations %>%
  dplyr::select(station) %>%
  dplyr::pull() %>%
  unique() %>%
  head()
```

The overall web design is specified, and this includes two selection inputs and one table output, specified by the function `tableOutput()`. This will allow us to display a tibble on our web page.

```
ui <- fluidPage(
  message("\nStarting the UI..."),
  titlePanel("Summarizing Monthly Rainfall"),
  selectInput("station1",
              label="Weather Station 1",
              choices=stations6),
  selectInput("station2",
              label="Weather Station 2",
              choices=stations6),
  tableOutput("monthly")
)
```

For the server code, just one output is set, and the function `renderTable()` is used to map back to the user interface, and will allow for the displaying of matrices and data frames. There are a number of steps in generating the required output.

- First, a filter operation is performed on the `observations` tibble based on the two selected inputs, which are indicated by the variables `input$station1` and `input$station2`.

- Next, the `dplyr` function `summarize()` is used to aggregate the monthly rainfall values, and these are transformed into a user-friendly format through the `tidyr` function `pivot_wider()`.

```
server <- function(input, output, session){

    message("\nStarting the server...")


    output$monthly <- renderTable({
        message(glue("output$table>>  with station1 = {input$station1}\n"))
        message(glue("output$table>>  with station2 = {input$station2}\n"))
        dataset <- filter(observations, station %in% c(input$station1,
                                                        input$station2))
        s_rain <- dataset %>%
                    dplyr::group_by(station, month) %>%
                    dplyr::summarize(TotalRain=sum(rain,na.rm = T)) %>%
                    tidyr::pivot_wider(names_from = "month",
                                        values_from="TotalRain")

        s_rain
    })
}
```

The app is then configured with the following call.

```
shinyApp(ui, server)
```

The app is then run, and an example of the output is shown in Figure 11.5. The advantage of widening the data can be seen, as the table provides a readable summary of the monthly rainfall data.

The reactive graph shows that any change in the two inputs ("station1" and "station2") will lead to a change in the output ("monthly"), where an updated version of the monthly rainfall values will be generated.

## 11.7   Example five: Creating a scatterplot

The aim of the fifth example is to show how plots can be rendered within the Shiny framework, and we return to the familiar example of the `ggplot2::mpg` dataset. We want to provide the user with the flexibility to select any two

(a) Shiny web page and components

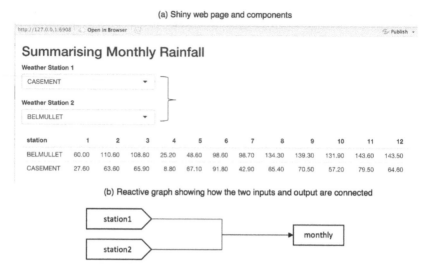

(b) Reactive graph showing how the two inputs and output are connected

**FIGURE 11.5** Summarizing montly rainfall for two stations

variables, and plot these on a scatterplot. To start the example, the relevant libraries are loaded.

```
library(shiny)
library(ggplot2)
library(glue)
```

Given that the tibble mpg is the target, we need to provide the user with a list of valid variables that can be used for the scatterplot. Any variable that is a numeric value can be included in this list (apart from the year), and we can use the following expression to extract all numeric columns names from mpg.

```
vars<- mpg %>%
        dplyr::select(-year) %>%
        dplyr::select(where(is.numeric)) %>%
        colnames()
```

Next, we construct the user interface. Within the fluidPage() function, there is an option to add a fluidRow(), which is a function that allows for the configuration of a row.

In this case, we add three components to the first row, two selection lists ("x_var" and "y_var"), and a checkbox component ("cb_smooth") which is used to specify whether a linear model visualization is required. There is one output component specified by a call to the function plotOutput(), which can then be a target of a renderPlot() call in the server function.

```
ui <- fluidPage(
  titlePanel("Exploring variables in the dataset mpg"),
  fluidRow(
    column(2,selectInput("x_var",
                          label="X Variable",
                          choices=vars)),
    column(2,selectInput("y_var",
                          label="Y Variable",
                          choices=vars)),
    column(5,checkboxInput("cb_smooth",
                          "Show Linear Model",
                          value = TRUE))
  ),
  plotOutput("plot")
)
```

The server function contains one function call, and this assigns the output of a call to `renderPlot()` to the variable `output$plot`. Inside the function call, we can observe the following sequence of code:

- The variables of interest (`x_var` and `y_var`) are extracted from two input components (`input$x_var` and `input$y_var`).
- The variable `p1` then stores the plot. Notice that we replace the usual call to `aes()` with a call to `aes_string()`. This is a useful function in `ggplot2` because we can pass in a string version of the variable and not the variable itself, which provides flexibility about which variable we want to select for the plot.
- If the logical value of the checkbox `input$cb_smooth` is `TRUE`, we will augment the plot variable `p1` with a call to `geom_smooth()`.
- The variable `p1` is returned, and this plot will then be rendered through the function `plotOutput()`.

```
server <- function(input, output, session){
  message("\nStarting the server...")
  output$plot <- renderPlot({
    x_var <- input$x_var
    y_var <- input$y_var
    message(glue("Smooth = {input$cb_smooth} \n"))
    p1 <- ggplot(mpg,aes_string(x=x_var,y=y_var))+
            geom_point()
    if(input$cb_smooth == TRUE)
      p1 <- p1 + geom_smooth(method="lm")
    p1
  })
}
```

We configure the app so that the user interface and server are connected.

```
shinyApp(ui, server)
```

The app is then run, and a sample of the output is shown in Figure 11.6. Notice the difference the call to `fluidRow()` has made, as the three input controls are all on one row of output, which results in a better usage of space. The output plot shows `displ` plotted against `hwy`, and it also highlights the linear model. Overall, an important feature of the Shiny plot is its flexibility, as the `ggplot2` function `aes_string()` allows any of the variables to be plotted, and therefore supports a wider range of user exploration.

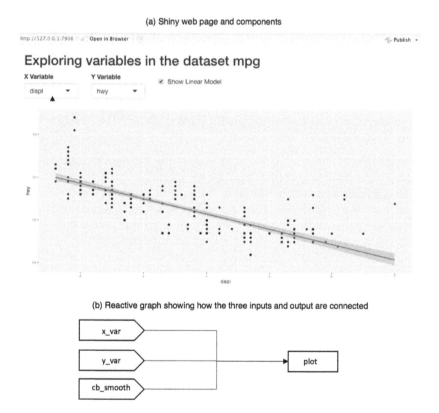

(a) Shiny web page and components

## Exploring variables in the dataset mpg

(b) Reactive graph showing how the three inputs and output are connected

**FIGURE 11.6** The web page produced for the Hello Shiny code

The reactive graph shows that a change in any one of the three inputs ("x_var", "y_var", or "cb_smooth") will lead to the generation of a new plot, which will be rendered to the web page. Therefore, even a simple action such as clicking on the checkbox will trigger an update to the page. This is because the code in the server that changes the variable `output$plot` directly references the variable `input$cb_smooth`. Therefore, any change in `input$cb_smooth` will trigger a response, or reaction, in `output$plot`.

## 11.8 Example six: Improving design by adding reactive expressions

Our final example introduces an additional feature of the Shiny framework, and it addresses a challenge that could arise when designing an interactive system. In this example, we introduce the idea of a *reactive expression* in Shiny, which allows the server to eliminate unnecessary computation when certain inputs change.

A reactive expression, defined with a call to the function `reactive()`, can make reference to a Shiny input, and then generate an output (i.e., the last evaluated expression). Adding reactive expressions can change the reactive graph, which provides information on how Shiny will generate output information based on what inputs change.

To illustrate how this process works, we will implement a new web page that generates random numbers from two Poisson distributions with means of $\lambda_1$ and $\lambda_2$, based on the total number of samples ($N$), and given a histogram bin width ($W$). The key learning point from this is to observe how the set of inputs can be decoupled from the server output, by using two new reactive expressions.

We first define the libraries to be used.

```
library(shiny)
library(ggplot2)
library(glue)
library(dplyr)
```

Next, we define the user interface. This consists of two rows. The first row contains four variables that comprise the required inputs, namely the means for the two Poisson distributions ($\lambda_1$ and $\lambda_2$), the number of samples to be drawn from the distributions ($N$), and the bin width of the histogram plot ($W$). The second row specifies the graphical output, and to create this output we will make use of `facet_wrap()` from ggplot2.

```
ui <- fluidPage(
  message("Starting the UI..."),
  titlePanel("Adding reactive expressions"),
  fluidRow(
  column(3,
         "Poisson One",
         numericInput("lambda1",
                      label="Lambda1",
                      value=50,
                      min=1)),
```

```
column(3,
        "Poisson Two",
        numericInput("lambda2",
                        label="Lambda2",
                        value=100,
                        min=1)),
column(3,
        "Total Samples",
        numericInput("N",
                        label="N",
                        value=1000,
                        min=100)),
column(3,
        "Binwidth",
        numericInput("BW",
                        label="W",
                        value=3,
                        min=1))),
fluidRow(column(12,plotOutput("plot")))
)
```

The server function contains three main elements, and two of these are *reactive expressions* which automatically cache their results, and only update when their respective inputs change (Wickham, 2021).

- The first reactive expression is where we define the variable `generate1` as the output from a call to the function `reactive()`. This reactive expression uses two Shiny inputs, the variables `input$N` and `input$lamdba1`, and generates a new tibble containing two variables, one for the distribution name ("Poisson One"), and the other containing each random number that has been generated via the function `rpois()`.

- The second reactive expression is where we define the variable `generate2` as a reactive expression. This uses two Shiny inputs, the variables `input$N` and `input$lamdba2`, and, similar to `generate1`, creates a new tibble containing the distribution name ("Poisson Two") and the random number from `rpois()`.

```
server <- function(input, output, session){
  message("\nStarting the server...")

  # Only gets called is input$N or input$lamda1 are changed
  generate1 <- reactive({
                    message("Calling generate1()...")
                    tibble(Distribution=rep("Poisson One",input$N),
                            Value=rpois(input$N,input$lambda1))})
```

```
# Only gets called is input$N or input$lamda2 are changed
generate2 <- reactive({
  message("Calling generate2()...")
  tibble(Distribution=rep("Poisson Two",input$N),
         Value=rpois(input$N,input$lambda2))})
```

```
output$plot <- renderPlot({
  message(glue("In output$graph >> {input$lambda1}"))
  message(glue("In output$graph >> {input$lambda2}"))
  d1 <- generate1()
  d2 <- generate2()
  d3 <- dplyr::bind_rows(d1, d2)
  ggplot(d3,aes(x=Value))+
      geom_histogram(binwidth = input$BW)+
      facet_wrap(~Distribution,nrow = 1,scales = "free")
})
```

```
}
```

- The final part of the server code is where we define the variable `output$plot`. This is where we can observe the impact of calling the two reactive expressions. Here, the variable `d1` contains the result from calling the reactive expression `generate1()`, while the variable `d2` contains the result returned by the reactive expression `generate2()`. The results from `d1` and `d2` are combined into `d3` using `dplyr::bind_rows()`, and the plot is returned showing both histograms, generated using the function `facet_wrap()`.

We configure the app so that the user interface and server are connected.

```
shinyApp(ui, server)
```

The web page inputs and outputs are shown in Figure 11.7. From a technical perspective, it is worth re-emphasizing that the reactive expressions cache their results, and only update these values when their inputs change. This means that the server code in `renderPlot()` will only execute `generate1()` when the input variables `input$N` or `input$lambda1` change, and that `generate2()` will only be reevaluated when either `input$N` or `input$lambda2` is modified. Therefore, if the bin width `input$BW` changes, the cached values from `generate1()` and `generate2()` will be used, and no new data will be sampled.

The reactive graph structure highlights the role the reactive expressions play *behind the scences* of a Shiny web page. It shows that the plotting component depends on inputs from `generate1()`, `generate2()`, and `input$BW`. The two reactive expressions will cache their outputs, and return these if their contributing inputs have not changed. If their inputs change, their outputs will be re-calculated. When the code is run, the output messages will confirm this,

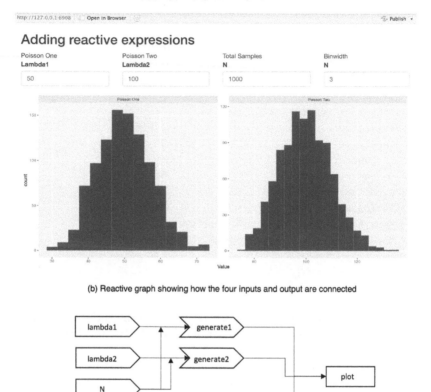

**FIGURE 11.7** Using reactive expression in Shiny

and will show which reactive expressions have been executed. The net benefit from this is greater efficiency in terms of leveraging the benefits of caching data that does not need to change.

This concludes the Shiny examples covered in the chapter, and the idea is to provide the reader with the basics of how to create an interactive app. Notice that in each example, only one output was considered, which is somewhat unrealistic given that it's likely that in an interactive web app there will be more than one output (e.g., plots, tables and text). For the interested reader, the textbook *Mastering Shiny* (Wickham, 2021) is recommended as a comprehensive text on building apps in Shiny, in particular, with its focus on mastering reactivity and best practice for implementation.

## 11.9    Summary of R functions from Chapter 11

A number of functions from the package shiny are now summarized, along with other functions used in the chapter.

| Function | Description |
|---|---|
| bind_rows() | Bind any number of data frames by row (dplyr). |
| fluidPage() | Creates fluid page layouts in Shiny. |
| glue() | Formats a string (library glue). |
| shinyApp() | Creates a Shiny app object from a UI/server pair. |
| message() | Generates a simple diagnostic message. |
| plotOutput() | Renders a renderPlot() within a web page. |
| numericInput() | Create a Shiny input control for numeric values. |
| reactive() | Creates a reactive expression. |
| renderTable() | Creates a reactive table that can display tibbles. |
| renderText() | Pastes output into a single string. |
| renderPlot() | Renders a reactive plot for an output. |
| renderPrint() | Prints the result of an expression. |
| selectInput() | Creates a selection list for user input. |
| tableOutput() | Creates a reactive table. Paired with renderTable(). |
| textOutput() | Render output as text. Paired with renderText(). |
| titlePanel() | Creates a panel containing an application title. |
| verbatimTextOutput() | Render a reactive output variable as text. |

## 11.10    Exercises

1. Draw a reactive graph from the following Shiny code.

```r
library(shiny)
library(glue)

ui <- fluidPage(
  numericInput("input_num1","Enter Number",10),
  numericInput("input_num2","Enter Number",10),
  textOutput("sum_msg"),
  textOutput("prod_msg")
)

server <- function(input, output, session){
```

```
output$sum_msg<- renderText({
  ans <- input$input_num1+input$input_num2
  glue("{input$input_num1} and {input$input_num2} is {ans}\n")
})

output$prod_msg<- renderText({
  ans <- input$input_num1*input$input_num2
  glue("{input$input_num1} times {input$input_num2} is {ans}\n")
})
}

shinyApp(ui, server)
```

2. Draw a reactive graph from the following Shiny code. Note that the code contains a reactive expression.

```
library(shiny)
library(glue)

ui <- fluidPage(
  numericInput("input_num1","Enter Number",10),
  numericInput("input_num2","Enter Number",10),
  verbatimTextOutput("diff_msg"),
  verbatimTextOutput("div_msg")
)

server <- function(input, output, session){

  msg <- reactive({
    ans <- input$input_num1-input$input_num2
    glue("{input$input_num1} minus {input$input_num2} is {ans}\n")
  })

  output$diff_msg<- renderText({
    msg();
  })

  output$div_msg<- renderText({
    ans <- input$input_num1/input$input_num2
    glue("{input$input_num1} divided by {input$input_num2} is {ans}\n")
  })
}

shinyApp(ui, server)
```

# Part III

# Using R for Operations Research

# 12

## Exploratory Data Analysis

> Exploratory analysis is what you do to understand the data and figure out what might be noteworthy or interesting to highlight to others. When we do exploratory analysis, it's like hunting for pearls in oysters.
>
> — Cole Nussbaumer Knaflic (Knaflic, 2015)

## 12.1 Introduction

Earlier in Part II, we presented tools from R's tidyverse, including the packages ggplot2 and dplyr, which facilitate efficient analysis of datasets. We now present an overall approach that provides a valuable guiding framework for initial analysis of a dataset. Exploratory data analysis (EDA) involves reviewing the features and characteristics of a dataset with an "open mind", and is frequently used upon "first contact with the data" (EDA, 2008). A convenient way to pursue EDA is to use questions as a means to guide your exploration, as this process focuses your attention on specific aspects of the dataset (Wickham et al., 2023). An attractive feature of EDA is that there are no constraints on the type of question posed, and therefore it can be viewed as a creative process. This chapter provides an overview of EDA, with a focus on five different datasets. The reader is encouraged to create their own questions that could drive an initial investigation of the data. Upon completing the chapter, you should have an appreciation for:

- The main idea underlying EDA, which is to ask questions of your dataset in an iterative way.
- How EDA can be applied to five CRAN datasets.
- How ggplot2 can be used to support EDA.

- How `dplyr` can be used to generate informative data summaries, and add new variables to datasets using the `mutate()` function.
- How the `ggpubr` function `stat_cor()` can be used to explore correlations between variables.
- The utility of the `lubridiate` package, and how it can generate additional time-related information from a timestamp variable.
- Related R functions that support exploratory data analysis in R.

### Chapter structure

- Section 12.2 presents an overview of an iterative process for EDA.
- Section 12.3 explores whether plant dimensions can help to identify a species.
- Section 12.4 explores the nature of the relationship between temperature and electricity demand.
- Section 12.5 investigates factors that may have impacted pupil–teacher ratios in Boston from the 1970s.
- Section 12.6 seeks to identify any patterns that increased a passenger's chance of survival on board the Titanic.
- Section 12.7 investigates the possible influence of wind direction on winter temperatures in Ireland.
- Section 12.8 provides a summary of the functions introduced in the chapter.

## 12.2    Exploratory data analysis

Many decision support and information systems development activities, for example operations research, simulation or software development, require an overall guiding method for problem solving. Typically these methods involve distinct phases, are iterative, and require a team effort. EDA is similar, and it is instructive to consider an overall organizing structure for the process of exploring datasets. Our approach is shown in Figure 12.1, and this is adapted from the exploratory data analysis workflow defined in the textbook *R for Data Science* (Wickham and Grolemund, 2016).

The stages within the EDA process are:

1. **Pose Question**, where an initial question is posed in order to direct focus on a particular aspect of a dataset. For example, a question could be (in relation to the tibble `ggplot2::mpg`), *do cars with big engines use more fuel than cars with small engines?* (Wickham and Grolemund, 2016).

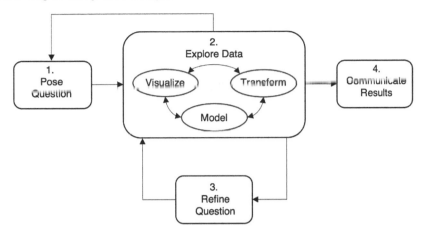

**FIGURE 12.1** A process for exploratory data analysis

2. **Explore Data**, where we can iterate through three activities to generate information that illuminates our question. These activities include:

    - *Visualize*, based on a target tibble, deploy the tools of `ggplot2` to visualize relevant variables and relationships.

    - *Transform*, where data from tibbles can be aggregated, joined, lengthened, or shortened. New variables can be added to data, for example, converting a date-time object to the weekday, something we will perform later in the chapter.

    - *Model*, where initial models can be built as part of the exploratory process. While our focus in this chapter is more on visualization and transformation, initial models can be shown using functions such as `geom_smooth()` and `stat_cor()`.

3. **Refine Question**, where based on insights generated (or not!), we can refine the question and start exploring the data again.

4. **Communicate Results**, where we can gather our insights into a presentation/report and communicate these to a wider audience. Tools such as RMarkdown (Xie et al., 2018) are invaluable, as they facilitate the creation of dynamic documents which can be reused to generate additional insights as the data is updated.

Here is an example of initial questions that we will explore in this chapter:

- Can a plant's dimensions help us uniquely identify a species?
- Based on observations from 2012, 2013, and 2104, is there a potential association between temperature and energy demand in the state of Victoria, Australia?

- For a Boston dataset from the 1970s, can we find possible relationships between house value and pupil–teacher ratios across different suburbs?
- What passengers had the greatest chance of survival on the Titanic?
- During the winter season in Ireland, is there initial evidence to suggest that wind direction has an impact on temperature?

## 12.3   Identifying species using plant measurements

Within the package datasets in R, the iris data frame is a collection of 150 observations based on data collected by the American botanist Edgar Anderson. For three species of the iris flower (setosa, versicolor, and virginica) measurements (in cm) relating to the length and width of both the sepal, and the petal are recorded. It is a popular dataset in the domain of data science, and can support modelling approaches such as decision trees and clustering. We can explore the tibble (note, we convert from a data frame for convenience) by selecting two observations from each species.

```
library(dplyr)
iris_tb <- dplyr::as_tibble(iris)
dplyr::slice(iris_tb,c(1:2,51:52,101:102))
#> # A tibble: 6 x 5
#>    Sepal.Length Sepal.Width Petal.Length Petal.Width Species
#>           <dbl>       <dbl>        <dbl>       <dbl> <fct>
#> 1          5.1         3.5          1.4         0.2 setosa
#> 2          4.9         3            1.4         0.2 setosa
#> 3          7           3.2          4.7         1.4 versicolor
#> 4          6.4         3.2          4.5         1.5 versicolor
#> 5          6.3         3.3          6           2.5 virginica
#> 6          5.8         2.7          5.1         1.9 virginica
```

As we have already discussed, exploratory data analysis involves starting with a question, and our question to explore is: *Can we identify the species of iris based on either the petal or sepal variables?*

Excluding the species variable, the initial dataset shows four columns for each observation: the sepal length and width, and the petal length and width. Our first task is to explore each of these continuous variables, and a histogram is a good choice for this, as it provides a valuable overall summary of each variable.

To simplify the visualization process, we can use the tidyr library to transform iris_tb into a simpler structure that has three variables: *Species*, which is the type of iris; *Measurement*, which represents the variable being measured, for example, the petal length or width; and *Value*, which contains the variable's value. The function pivot_longer() is used the generate the transformed tibble.

```
library(tidyr)
iris_long <- tidyr::pivot_longer(iris_tb,
                                 names_to = "Measurement",
                                 values_to = "Value",
                                 -Species)
head(iris_long)
#> # A tibble: 6 x 3
#>   Species Measurement  Value
#>   <fct>   <chr>        <dbl>
#> 1 setosa  Sepal.Length  5.1
#> 2 setosa  Sepal.Width   3.5
#> 3 setosa  Petal.Length  1.4
#> 4 setosa  Petal.Width   0.2
#> 5 setosa  Sepal.Length  4.9
#> 6 setosa  Sepal.Width   3
```

Histograms are generated from the following R code, and the resulting plot is visualized in Figure 12.2. Note the `alpha` argument controls the geom's opacity, and we fill the histogram bars based on the species.

```
p1 <- ggplot(iris_long,aes(x=Value,fill=Species))+
      geom_histogram(color="white", alpha=0.7)+
      facet_wrap(~Measurement,ncol=2)+
      theme(legend.position = "top")
p1
#> `stat_bin()` using `bins = 30`. Pick better value with
#> `binwidth`.
```

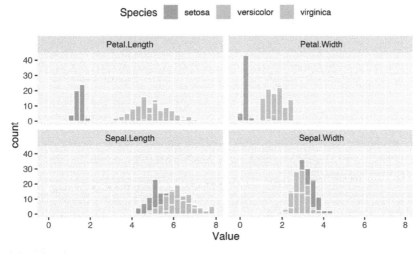

**FIGURE 12.2** Exploring the iris species data

The histograms provide insights into the data. For example, the petal variables visualized on the top row show a clearer separation between the species. Specifically, for both the petal length and width, the setosa histogram plot is fully distinct from the other two. There also seems to be a difference between the versicolor and virginica species.

Using visual inspection, the petal attributes provide a clearer differentiation between the species. We can now extract summary values. Of course, these summaries could also be obtained by simply calling the `summary(iris_tb)`, however, the advantage here is that we create a tibble with this overall information where all the values can be easily compared. We combine the tools of `dplyr` to generate the following summary, which is stored in the tibble `res`.

```
res <- iris_long %>%
        dplyr::filter(Measurement %in% c("Petal.Length",
                                         "Petal.Width")) %>%
        dplyr::group_by(Species,Measurement) %>%
        dplyr::summarize(Min=min(Value),
                         Q25=quantile(Value,0.25),
                         Median=median(Value),
                         Mean=mean(Value),
                         Q75=quantile(Value,0.75),
                         Max=max(Value)) %>%
        dplyr::ungroup() %>%
        dplyr::arrange(Measurement,Mean)
res
#> # A tibble: 6 x 8
#>   Species    Measurement    Min   Q25 Median  Mean   Q75   Max
#>   <fct>      <chr>        <dbl> <dbl>  <dbl> <dbl> <dbl> <dbl>
#> 1 setosa     Petal.Length     1   1.4   1.5  1.46  1.58   1.9
#> 2 versicolor Petal.Length     3     4  4.35  4.26   4.6   5.1
#> 3 virginica  Petal.Length   4.5   5.1  5.55  5.55  5.88   6.9
#> 4 setosa     Petal.Width    0.1   0.2   0.2 0.246   0.3   0.6
#> 5 versicolor Petal.Width      1   1.2   1.3  1.33   1.5   1.8
#> 6 virginica  Petal.Width    1.4   1.8     2  2.03   2.3   2.5
```

The data provides further insights, as the setosa iris has the lowest mean value, while virginica has the largest mean. The other values provide additional information, and would suggest that both variables could be used to help classify the iris species.

To confirm this, in an informal way, we can return to the original tibble and generate a scatterplot using the petal length and width, and color the dots by the species. The resulting plot is shown in Figure 12.3.

```
p2 <- ggplot(iris_tb,aes(x=Petal.Length,y=Petal.Width,color=Species)) +
```

```
geom_point()
```
p2

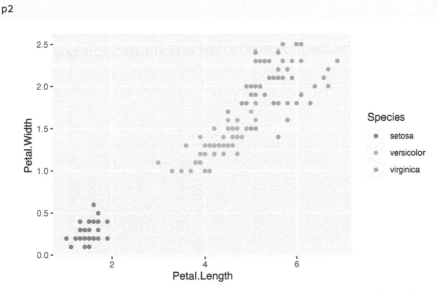

FIGURE 12.3 Exploring the association between petal length and petal width

The output reveals an interesting pattern which shows how the scatterplot petal length and petal width leads shows potential clustering of the different species. These relationships can be explored more fully, for example, by modelling the data using a decision tree classifier, which can generate rules that classify inputs (e.g., petal length and width) into outputs (one of the three species of iris) (Lantz, 2019). This modelling is outside the chapter's scope, but would fall within the exploratory data analysis process, and the interested reader is encouraged to further explore the dataset using techniques such as decision trees and unsupervised learning.

## 12.4 Exploring electricity demand in Victoria, Australia

Our next example focuses on time series data recorded in Victoria, Australia from 2012 through to 2014, and it can be accessed through the CRAN tibble tsibbledata (Hyndman and Athanasopoulos, 2021). This package is part of a suite of packages that provide excellent support for forecasting and time series analysis methods in R, and this dataset contains electricity demand values (megawatt hours) and the corresponding temperature levels (celsius) at 30 minute intervals. Because of this, the dataset supports exploratory analysis of (1) the associations between temperature and electricity demand, and (2) patterns of electricity demand by time of the year (e.g., quarters).

To load and prepare the data, we need to include a number of extra libraries, as shown below.

```
library(ggplot2)
library(dplyr)
library(tsibble)
library(tsibbledata)
library(lubridate)
library(tidyr)
library(ggpubr)
```

The library `tsibble` specifies a new type of tibble that has been created to support time series analysis (which is outside of this chapter's scope). In this example, the variable `vic_elec` is a `tsibble`, and we can see its S3 structure by calling `class()`, where it is clearly sub-classed from both tibbles and a data frame.

```
class(vic_elec)
#> [1] "tbl_ts"     "tbl_df"     "tbl"         "data.frame"
```

However, for our purposes, because we are not performing any time series analysis, we will recast the object to a tibble, and use the function `tsibble::as_tibble()` to perform this task. We create a new variable `aus_elec`, and we can observe its structure below.

```
aus_elec <- vic_elec %>%
              tsibble::as_tibble()
aus_elec %>% dplyr::slice(1:3)
#> # A tibble: 3 x 5
#>    Time                 Demand Temperature Date        Holiday
#>    <dttm>                <dbl>       <dbl> <date>      <lgl>
#> 1 2012-01-01 00:00:00   4383.        21.4 2012-01-01 TRUE
#> 2 2012-01-01 00:30:00   4263.        21.0 2012-01-01 TRUE
#> 3 2012-01-01 01:00:00   4049.        20.7 2012-01-01 TRUE
```

There are four variables in `aus_elec`:

- `Time`, which is a timestamp of the observation time. This is an important R data structure, as from this we can extract, using the package `lubridate`, addition time-related features such as the weekday.
- `Demand`, the total electricity demand in megawatt hours (MWh).
- `Temperature`, the temperature in Melbourne, which is the capital of the Australian state of Victoria.
- `Date`, the date of each observation.
- `Holiday`, a logical value indicating whether the day is a public holiday.

As part of our exploratory data analysis, we want to create additional variables based on the variable Time. These can be extracted using functions from the package lubridate, for example:

- wday(), which returns the weekday and can be represented as a factor (e.g., Sun, Mon, etc.).
- year(), which returns the year.
- month(), which returns the month.
- hour(), which returns the hour.
- yday(), which returns the day number for the year (1 to 365 or 366).

With these features, we create two additional variables: (1) the quarter of the year to capture possible seasonal flucuations, where this value is based on the month; and, (2) the day segment, where each day is divided into four quarters, and this calculation is based on the hour of the day. Note the use of the dplyr function case_when() to create the variables. Overall, as we are adding new variables, we use the dplyr function mutate(), and then we show the first three observations.

```r
aus_elec <- aus_elec %>%
            dplyr::mutate(WDay=wday(Time,label=TRUE),
                    Year=year(Time),
                    Month=as.integer(month(Time)),
                    Hour=hour(Time),
                    YearDay=yday(Time),
                    Quarter=case_when(
                      Month %in% 1:3   ~ "Q1",
                      Month %in% 4:6   ~ "Q2",
                      Month %in% 7:9   ~ "Q3",
                      Month %in% 10:12 ~ "Q4",
                      TRUE ~ "Undefined"
                    ),
                    DaySegment=case_when(
                      Hour %in%  0:5  ~ "S1",
                      Hour %in%  6:11 ~ "S2",
                      Hour %in% 12:17 ~ "S3",
                      Hour %in% 18:23 ~ "S4",
                      TRUE ~ "Undefined"
                    ))
```

```r
dplyr::slice(aus_elec,1:3)
#> # A tibble: 3 x 12
#>    Time                Demand Temperature Date       Holiday WDay
#>    <dttm>               <dbl>       <dbl> <date>     <lgl>   <ord>
#> 1 2012-01-01 00:00:00   4383.        21.4 2012-01-01 TRUE    Sun
#> 2 2012-01-01 00:30:00   4263.        21.0 2012-01-01 TRUE    Sun
```

```
#> 3 2012-01-01 01:00:00   4049.            20.7 2012-01-01 TRUE      Sun
#> # ... with 6 more variables: Year <dbl>, Month <int>,
#> #   Hour <int>, YearDay <dbl>, Quarter <chr>, DaySegment <chr>
```

We can now create our first visualization of the dataset and use pivot_longer()
to generate a flexible data structure containing the Time, Quarter, Indicator,
and Value.

```
aus_long <- aus_elec %>%
           dplyr::select(Time,Demand,Temperature,Quarter) %>%
           tidyr::pivot_longer(names_to="Indicator",
                               values_to="Value",
                               -c(Time,Quarter))
dplyr::slice(aus_long,1:3)
#> # A tibble: 3 x 4
#>    Time                 Quarter Indicator    Value
#>    <dttm>               <chr>   <chr>        <dbl>
#> 1 2012-01-01 00:00:00 Q1      Demand       4383.
#> 2 2012-01-01 00:00:00 Q1      Temperature  21.4
#> 3 2012-01-01 00:30:00 Q1      Demand       4263.
```

```
p3 <- ggplot(aus_long,aes(x=Time,y=Value,color=Quarter))+
      geom_point(size=0.2)+
      facet_wrap(~Indicator,scales="free",ncol = 1)+
      theme(legend.position = "top")
p3
```

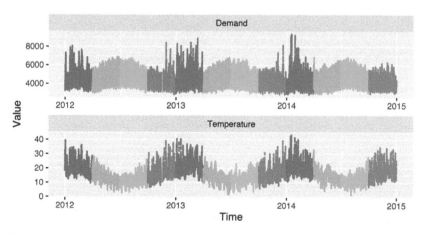

**FIGURE 12.4** Time series of temperature and electricity demand (by quarter)

With this data we create two plots, shown in Figure 12.4, one for temperature (bottom plot), and the other for electricity demand (top plot), where the points are colored by quarter. It is interesting to explore possible associations between the temperature and demand, for the different quarters of the year. For example, it does seem that the quarter one values for temperature and demand seem to move in the same direction. To further explore this, we can calculate the correlation coefficient between temperature and demand for each quarter of every year.

We revert to the dataset `aus_elec`, as this contains both the temperature and electricity demand for each quarter. Using `dplyr` functions, we can calculate the correlation coefficient for each year and quarter, and sort from highest correlation value to lowest.

```
aus_cor <- aus_elec %>%
          dplyr::group_by(Year, Quarter) %>%
          dplyr::summarize(CorrCoeff=cor(Temperature, Demand)) %>%
          dplyr::ungroup() %>%
          dplyr::arrange(desc(CorrCoeff))
#> `summarize()` has grouped output by 'Year'. You can override
#> using the `.groups` argument.
dplyr::slice(aus_cor,1:4)
#> # A tibble: 4 x 3
#>     Year Quarter CorrCoeff
#>    <dbl> <chr>       <dbl>
#> 1   2014 Q1          0.796
#> 2   2013 Q1          0.786
#> 3   2012 Q1          0.683
#> 4   2012 Q4          0.476
```

Interestingly, the highest values are from quarter one (January, February, and March), which are also those months with high overall temperature values.

It may also be informative to layer the actual correlation coefficient data onto a plot of Demand v Temperature. The function `stat_cor()`, which is part of the package `ggpubr`, can be used for this, as it displays both the correlation coefficients and the associated p-values. We show this output in Figure 12.5, for the first and third quarters.

```
q1_out <- filter(aus_elec, Quarter %in% c("Q1","Q3"))
p4 <- ggplot(q1_out,aes(x=Temperature,y=Demand))+
      geom_point(alpha=0.2,size=0.5)+
      facet_grid(Year~Quarter)+
      stat_cor(digits = 2)
p4
```

What is helpful about this visualization is that quarterly values (for "Q1" and "Q3") are contained in each column, and so they can be easily compared

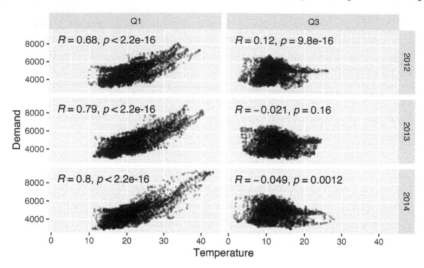

**FIGURE 12.5** Associations between temperature and demand, by year and quarter

across the 3 years. It shows high correlation values for the first quarter, when temperatures are high. The third quarter has low correlation values, due to the lower temperature levels between July and September.

Our next analysis is to explore each quarter for every year, and extract the maximum electricity demand for the quarter. Interestingly, we will also extract additional variables for the time of the maximum demand, including the temperature, day of week, and day segment. This is achieved by combining two different functions. First, the function which.max() will extract the row index for the maximum observation for Demand. Second, this index value (stored in MaxIndex) is an input argument to the function nth(), which returns the appropriate value. We order the values from highest to lowest demand.

```
aus_summ <- aus_elec %>%
            dplyr::group_by(Year,Quarter) %>%
            dplyr::summarize(MaxD=max(Demand),
                             MaxIndex=which.max(Demand),
                             Time=nth(Time,MaxIndex),
                             Temp=nth(Temperature,MaxIndex),
                             Day=nth(WDay,MaxIndex),
                             DaySegment=nth(DaySegment,MaxIndex)) %>%
            dplyr::ungroup() %>%
            dplyr::arrange(desc(MaxD))
head(aus_summ)
#> # A tibble: 6 x 8
#>     Year Quarter  MaxD MaxIndex Time                  Temp Day
```

```
#>      <dbl> <chr>   <dbl>    <int> <dttm>              <dbl> <ord>
#> 1    2014 Q1      9345.      755 2014-01-16 17:00:00  38.8 Thu
#> 2    2013 Q1      8897.     3395 2013-03-12 17:00:00  35.5 Tue
#> 3    2012 Q4      8443.     2865 2012-11-29 17:00:00  38.7 Thu
#> 4    2013 Q4      8138.     3824 2013-12-19 16:30:00  39   Thu
#> 5    2012 Q1      8072.     1138 2012-01-24 16:30:00  33.6 Tue
#> 6    2012 Q2      6921.     3926 2012-06-21 17:30:00   9.6 Thu
#> # ... with 1 more variable: DaySegment <chr>
```

The output from this query can be cross-checked against a specific filtering call on the tibble. For example, the highest recorded value is for quarter Q1 and year 2014, and it shows that the index for this record was row 755. We can then extract this directly from the tibble using the following code, and this result confirms that the observation is the same.

```
aus_elec %>%
  dplyr::filter(Quarter=="Q1",Year==2014) %>%
  dplyr::slice(755)
#> # A tibble: 1 x 12
#>    Time                Demand Temperature Date       Holiday WDay
#>    <dttm>               <dbl>       <dbl> <date>       <lgl> <ord>
#> 1 2014-01-16 17:00:00  9345.        38.8 2014-01-16 FALSE    Thu
#> # ... with 6 more variables: Year <dbl>, Month <int>,
#> #   Hour <int>, YearDay <dbl>, Quarter <chr>, DaySegment <chr>
```

For further exploratory analysis of this dataset, the reader is encouraged to explore the forecasting textbook (Hyndman and Athanasopoulos, 2021), as it also includes excellent visualization functions for time series data, and access to a suite of forecasting methods.

## 12.5 Exploring housing values in the Boston suburbs

This dataset is accessed from the CRAN package MASS, and it contains information on Boston suburb housing prices, and related variables, from the 1970s (Harrison Jr and Rubinfeld, 1978). The dataset contains 14 variables, and for the purpose of our exploratory analysis, we focus on the following:

- chas, a variable that indicates whether the area bounds the Charles River (1 = bounds, 0 otherwise).
- rm, the average number of rooms per dwelling.
- age, the proportion of owner-occupied units built prior to 1940.
- rad, index of accessibility to radial highways.
- ptratio, pupil–teacher ratio by suburb.

- `medv`, the median value of owner-occupied homes (thousands of dollars).
- `nox`, the nitrogen oxide concentration (in parts per million).

We first load the packages that are required for exploratory data analysis. There is a function conflict issue here, as a `select()` function is defined in the packages MASS and dplyr (used for column selection). To address this, we will use the full function call `dplyr::select()` to pick columns from Boston.

```
library(MASS)
library(ggplot2)
library(GGally)
library(dplyr)
library(tidyr)
```

The Boston dataset columns can be viewed.

```
dplyr::glimpse(Boston)
#> Rows: 506
#> Columns: 14
#> $ crim    <dbl> 0.00632, 0.02731, 0.02729, 0.03237, 0.06905, 0.~
#> $ zn      <dbl> 18.0, 0.0, 0.0, 0.0, 0.0, 0.0, 12.5, 12.5, 12.5~
#> $ indus   <dbl> 2.31, 7.07, 7.07, 2.18, 2.18, 2.18, 7.87, 7.87,~
#> $ chas    <int> 0, 0, 0, 0, 0, 0, 0, 0, 0, 0, 0, 0, 0, 0, 0, 0,~
#> $ nox     <dbl> 0.538, 0.469, 0.469, 0.458, 0.458, 0.458, 0.524~
#> $ rm      <dbl> 6.575, 6.421, 7.185, 6.998, 7.147, 6.430, 6.012~
#> $ age     <dbl> 65.2, 78.9, 61.1, 45.8, 54.2, 58.7, 66.6, 96.1,~
#> $ dis     <dbl> 4.090, 4.967, 4.967, 6.062, 6.062, 6.062, 5.561~
#> $ rad     <int> 1, 2, 2, 3, 3, 3, 5, 5, 5, 5, 5, 5, 5, 4, 4, 4,~
#> $ tax     <dbl> 296, 242, 242, 222, 222, 222, 311, 311, 311, 31~
#> $ ptratio <dbl> 15.3, 17.8, 17.8, 18.7, 18.7, 18.7, 15.2, 15.2,~
#> $ black   <dbl> 396.9, 396.9, 392.8, 394.6, 396.9, 394.1, 395.6~
#> $ lstat   <dbl> 4.98, 9.14, 4.03, 2.94, 5.33, 5.21, 12.43, 19.1~
#> $ medv    <dbl> 24.0, 21.6, 34.7, 33.4, 36.2, 28.7, 22.9, 27.1,~
```

However, for our initial analysis, we focus on the seven variables previously outlined, and we will grab this opportunity to use the dplyr function `rename()` to create more descriptive titles for these variables. This is a common task in data science, as we may want to alter column names to suit our purpose. In this code, we also convert the data frame to a tibble, and set the Charles River variable to a logical value. The resulting data is stored in the tibble bos.

```
bos <- Boston %>%
       dplyr::as_tibble() %>%
       dplyr::select(chas,rm,age,rad,ptratio,medv,nox) %>%
       dplyr::rename(PTRatio=ptratio,
                     ByRiver=chas,
                     Rooms=rm,
                     Age=age,
```

```
                        Radial=rad,
                        Value=medv,
                        Nox=nox) %>%
        dplyr::mutate(ByRiver=as.logical(ByRiver))
bos
#> # A tibble: 506 x 7
#>    ByRiver Rooms   Age Radial PTRatio Value    Nox
#>    <lgl>   <dbl> <dbl>  <int>   <dbl> <dbl>  <dbl>
#>  1 FALSE    6.58  65.2      1    15.3  24    0.538
#>  2 FALSE    6.42  78.9      2    17.8  21.6  0.469
#>  3 FALSE    7.18  61.1      2    17.8  34.7  0.469
#>  4 FALSE    7.00  45.8      3    18.7  33.4  0.458
#>  5 FALSE    7.15  54.2      3    18.7  36.2  0.458
#>  6 FALSE    6.43  58.7      3    18.7  28.7  0.458
#>  7 FALSE    6.01  66.6      5    15.2  22.9  0.524
#>  8 FALSE    6.17  96.1      5    15.2  27.1  0.524
#>  9 FALSE    5.63 100        5    15.2  16.5  0.524
#> 10 FALSE    6.00  85.9      5    15.2  18.9  0.524
#> # ... with 496 more rows
```

Our initial question is to explore the pupil–teacher ratio and observe how other variables might be correlated with this measure. We can visualize relationships between continuous variables using the function ggpairs(), and we use the following code to explore the data. Note that here we exclude the categorical variable ByRiver from this analysis, and the argument progress=FALSE excludes progress bar during the plotting process.

```
p3 <- ggpairs(dplyr::select(bos,-ByRiver),progress = FALSE)+
  theme_light()+
  theme(axis.text.x = element_text(size=6),
        axis.text.y = element_text(size=6))
p3
```

This plot, shown in Figure 12.6, provides three types of statistical information.

- The diagonal visualizes the density plot for each variable; in this case, the number of rooms seems the closest representation of a normal distribution.

- The lower triangular section of the plot (excluding the diagonal) presents a scatterplot of the relevant row and column variable. These values can be quickly viewed to see possible linear relationships. In this case, location *row 4, column 1* on the plot shows a strong positive relationship between Value and Rooms.

- The upper triangular section (again excluding the diagonal) shows the correlation coefficient between the row and column variables, and so it complements the related plot in the lower triangular section. Here we see, in location *row*

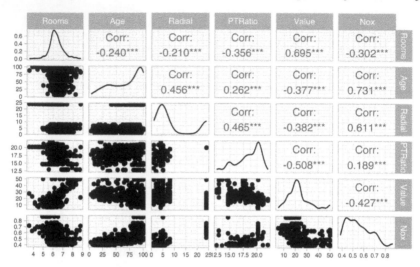

**FIGURE 12.6** Exploring pair-wise associations in the Boston dataset

*1, column 4*, that the correlation coefficient between `Value` and `Room` is 0.695, indicating that these variables are positively correlated.

Of course, the R function `cor` can also be directly used to summarize these correlations.

```
cor(dplyr::select(bos,-ByRiver))
#>              Rooms      Age  Radial PTRatio    Value     Nox
#> Rooms       1.0000  -0.2403 -0.2098 -0.3555   0.6954 -0.3022
#> Age        -0.2403   1.0000  0.4560  0.2615  -0.3770  0.7315
#> Radial     -0.2098   0.4560  1.0000  0.4647  -0.3816  0.6114
#> PTRatio    -0.3555   0.2615  0.4647  1.0000  -0.5078  0.1889
#> Value       0.6954  -0.3770 -0.3816 -0.5078   1.0000 -0.4273
#> Nox        -0.3022   0.7315  0.6114  0.1889  -0.4273  1.0000
```

From our initial question relating to the pupil–teacher ratio, we can observe that the strongest correlation between this and the other variables is –0.508, for the variable `Value`. This shows that the variables are negatively correlated, hence an increase in the `Value` is associated with a decrease in `PTRatio`.

An interesting feature of the dataset is the inclusion of a categorical variable `ByRiver` which indicates whether a suburb intersects with the Charles River. An initial comparison of all the continuous variables can be made, where the summary statistics can be presented for the `ByRiver` variable. Note that in the original dataset this value was an integer (1 or 0), but in creating our dataset, we converted this to a logical value (`TRUE` for bordering the river, `FALSE` for not bordering the river). To prepare the data for this new visualization, we utilize `pivot_longer()`, which reduces the number of columns to three.

```
bos_long <- bos %>%
              tidyr::pivot_longer(names_to = "Indicator",
                                  values_to = "Value",
                                  -ByRiver)
bos_long
#> # A tibble: 3,036 x 3
#>     ByRiver Indicator   Value
#>     <lgl>   <chr>       <dbl>
#>   1 FALSE   Rooms        6.58
#>   2 FALSE   Age         65.2
#>   3 FALSE   Radial       1
#>   4 FALSE   PTRatio     15.3
#>   5 FALSE   Value       24
#>   6 FALSE   Nox          0.538
#>   7 FALSE   Rooms        6.42
#>   8 FALSE   Age         78.9
#>   9 FALSE   Radial       2
#>  10 FALSE   PTRatio     17.8
#> # ... with 3,026 more rows
```

With this longer tibble, we generate a boxplot for each variable, and color these plots with the variable ByRiver. A facet plot is created using the variable Indicator, which allows a single plot per variable. The code for generating the plot is shown below, and this is followed by the plot itself, which is visualized in Figure 12.7.

```
p5 <- ggplot(bos_long,aes(x=ByRiver,y=Value,color=ByRiver))+
        geom_boxplot()+
        facet_wrap(~Indicator,scales="free")+
        theme(legend.position = "top")
p5
```

This visualization contributes further to our analysis of the initial question. It shows that houses that are bordering the Charles River have a lower median pupil–teacher ratio than other houses. However, a further clue to the reason for this can be gleaned from observing the two boxplots for Value, which show a higher median for those houses close to the river. Therefore, as Value was already shown to be negatively correlated with PTRatio, it makes sense that the median of PTRatio is lower in suburbs closer to the river.

In summary, this dataset from MASS supports exploratory data analysis, and other statisical methods (which are outside of this book's scope) such as analysis of variance (ANOVA) (Crawley, 2015) can be used to explore whether these differences between variables are statistically significant.

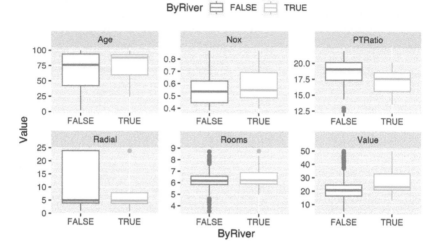

**FIGURE 12.7** Box plots of indicators categorized by proximity to Charles River

## 12.6   Exploring passenger survival chances on board the Titanic

The Titanic was built in Belfast, and departed on its first voyage from Southampton on April 10, 1912, stopping in Cherbourg and Cobh, before setting sail across the Atlantic Ocean on route to New York. On April 15, she struck an iceberg off the coast of Canada, and sank to the ocean floor. Over 1,500 passengers and crew perished, with only about 700 surviving. Electronic versions of the ship's passenger list are now available, and have been used as case studies for statistical analysis, and machine learning, in order to explore survival outcomes of passengers. In this section, we use a version of the passenger dataset provided by the CRAN package titanic, which contains a sample of 891 passengers within the tibble titanic_train, along with 12 features for each individual. (Note there is another tibble within this package called titanic_test which contains information on additional passengers, but this is not used in our example.)

To load the data, we include the relevant tidyverse libraries, along with the titanic library, and then we provide a summary of the tibble titanic_train, using the dplyr function glimpse().

```
library(ggplot2)
library(dplyr)
library(titanic)
```

```
dplyr::glimpse(titanic_train)
#> Rows: 891
#> Columns: 12
#> $ PassengerId <int> 1, 2, 3, 4, 5, 6, 7, 8, 9, 10, 11, 12, 13, ~
#> $ Survived    <int> 0, 1, 1, 1, 0, 0, 0, 0, 1, 1, 1, 1, 0, 0, 0~
#> $ Pclass      <int> 3, 1, 3, 1, 3, 3, 1, 3, 3, 2, 3, 1, 3, 3, 3~
#> $ Name        <chr> "Braund, Mr. Owen Harris", "Cumings, Mrs. J~
#> $ Sex         <chr> "male", "female", "female", "female", "male~
#> $ Age         <dbl> 22, 38, 26, 35, 35, NA, 54, 2, 27, 14, 4, 5~
#> $ SibSp       <int> 1, 1, 0, 1, 0, 0, 0, 3, 0, 1, 1, 0, 0, 1, 0~
#> $ Parch       <int> 0, 0, 0, 0, 0, 0, 0, 1, 2, 0, 1, 0, 0, 5, 0~
#> $ Ticket      <chr> "A/5 21171", "PC 17599", "STON/O2. 3101282"~
#> $ Fare        <dbl> 7.250, 71.283, 7.925, 53.100, 8.050, 8.458,~
#> $ Cabin       <chr> "", "C85", "", "C123", "", "", "E46", "", "~
#> $ Embarked    <chr> "S", "C", "S", "S", "S", "Q", "S", "S", "S"~
```

The idea behind our exploratory data analysis is to see which of these factors were important for survival. To prepare our dataset, we simplify our focus to five variables:

- PassengerID, a unique number for each passenger.
- Survived, the survival indicator (1=survived, 0=perished).
- Pclass, the passenger class (1=first, 2=second, and 3=third).
- Sex, passenger sex ("male" or "female").
- Age, age of passenger.

For convenience, as part of the data preparation process, we transform the variable Survived to a logical value, set the Pclass variable to a factor, and also factorize the variable Sex. This will provide informative summaries of our data, as we demonstrate below.

```
titanic <- titanic_train %>%
            dplyr::select(PassengerId,
                    Survived,
                    Pclass,
                    Sex,
                    Age) %>%
            dplyr::mutate(Survived=as.logical(Survived),
                    Sex=factor(Sex,
                            levels=c("male","female")),
                    Class=factor(Pclass,
                            levels=c(1,2,3))) %>%
            dplyr::select(-Pclass) %>%
            dplyr::as_tibble()
summary(titanic)
#>    PassengerId    Survived            Sex             Age
```

```
#>   Min.    :  1     Mode :logical     male   :577     Min.    : 0.42
#>   1st Qu.:224     FALSE:549         female:314     1st Qu.:20.12
#>   Median :446     TRUE :342                         Median :28.00
#>   Mean    :446                                       Mean    :29.70
#>   3rd Qu.:668                                       3rd Qu.:38.00
#>   Max.    :891                                       Max.    :80.00
#>                                                      NA's    :177
#>   Class
#>   1:216
#>   2:184
#>   3:491
#>
#>
#>
#>
```

This also highlights the utility of the `summary()` function. For continuous variables, a six variable summary is generated, and for categorical variables, a count of observations for each factor level is displayed.

Our question to explore relates to an analysis of the survival data, and as a starting point, we can generate a high-level summary of the overall survival rates and store this as the variable `sum1`. This includes the total number of observations (`N`), the total who survived (`TSurvived`), the total who perished (`TPerished`), and the proportion who survived and perished. Note we take advantage of the fact that the `summarize()` function can use its calculated values in new column calculations, for example, `PropSurvived` uses the previously calculated value of `N`. This code also shows that the `summarize()` can be used on the complete dataset, although, as we have seen throughout the book, we mostly deploy it with the function `group_by()`.

```
sum1 <- titanic %>%
        dplyr::summarize(N=n(),
                         TSurvived=sum(Survived),
                         TPerished=sum(Survived==FALSE),
                         PropSurvived=TSurvived/N,
                         PropPerished=TPerished/N)
sum1
#> # A tibble: 1 x 5
#>       N TSurvived TPerished PropSurvived PropPerished
#>   <int>    <int>    <int>        <dbl>        <dbl>
#> 1   891      342      549        0.384        0.616
```

The main result here is that just over 38% of the sample survived; however, the survival rates of the different subgroups are not clear. The next task is to drill down into the data and explore the same variables to see whether there are differences by sex. This is a straightforward task, as we simply add the command `group_by()` into the data processing pipeline and arrange them in decreasing order (by proportion survived). We can then observe the results, which are stored in `sum2`.

```
sum2 <- titanic %>%
          dplyr::group_by(Sex) %>%
          dplyr::summarize(N=n(),
                      TSurvived=sum(Survived),
                      TPerished=sum(Survived==FALSE),
                      PropSurvived=TSurvived/N,
                      PropPerished=TPerished/N) %>%
                      arrange(desc(PropSurvived))
sum2
#> # A tibble: 2 x 6
#>    Sex        N TSurvived TPerished PropSurvived PropPerished
#>    <fct>  <int>     <int>     <int>        <dbl>        <dbl>
#> 1 female   314       233        81        0.742        0.258
#> 2 male     577       109       468        0.189        0.811
```

A related query is to explore this data with respect to the passenger's class. Again, it's a straightforward addition to the high-level query, where we add a `group_by()` function call to the overall data transformation pipeline. Similar to the previous command, the results are sorted by `PropSurvived` and the overall results are stored in the tibble `sum3`.

```
sum3 <- titanic %>%
          dplyr::group_by(Class) %>%
          dplyr::summarize(N=n(),
                      TSurvived=sum(Survived),
                      TPerished=sum(Survived==FALSE),
                      PropSurvived=TSurvived/N,
                      PropPerished=TPerished/N) %>%
          dplyr::arrange(desc(PropSurvived))
sum3
#> # A tibble: 3 x 6
#>    Class     N TSurvived TPerished PropSurvived PropPerished
#>    <fct> <int>     <int>     <int>        <dbl>        <dbl>
#> 1 1       216       136        80        0.630        0.370
#> 2 2       184        87        97        0.473        0.527
#> 3 3       491       119       372        0.242        0.758
```

Here we observe a difference in the groups, with the highest survival rate proportion (at ~63%) for those in first class, while the lowest value (~24%) are those who purchased third-class tickets.

Our next summary groups the data by the two variables to generate another set of results, this time, for each sex and for each travel class. Again, the flexibility of the `group_by()` function is evident, as we combine these two variables, which will then generate six columns, one for each combination of variables. The results are stored in the tibble `sum4`, and once more, we arrange these in descending order of the variable `PropSurvived`.

```
sum4 <- titanic %>%
        dplyr::group_by(Class,Sex) %>%
        dplyr::summarize(N=n(),
                         TSurvived=sum(Survived),
                         TPerished=sum(Survived==FALSE),
                         PropSurvived=TSurvived/N,
                         PropPerished=TPerished/N) %>%
        dplyr::arrange(desc(PropSurvived))
sum4
#> # A tibble: 6 x 7
#> # Groups:    Class [3]
#>   Class Sex        N TSurvived TPerished PropSurvived PropPeris~1
#>   <fct> <fct>  <int>     <int>     <int>        <dbl>       <dbl>
#> 1 1     female    94        91         3        0.968      0.0319
#> 2 2     female    76        70         6        0.921      0.0789
#> 3 3     female   144        72        72        0.5        0.5
#> 4 1     male     122        45        77        0.369      0.631
#> 5 2     male     108        17        91        0.157      0.843
#> 6 3     male     347        47       300        0.135      0.865
#> # ... with abbreviated variable name 1: PropPerished
```

This data transformation provides additional insights, specifically, that the survival proportions for females in first and second class were each over 90%, and the lowest survival proportions were males in second and third class. This shows the differences in outcome based on the two categorical variables, `Sex` and `Class`. We can now provide a fresh perspective on these results by visually representing the data. The following code presents the results, where we show a bar chart for the variable `Survived`, and we display this for the two categories of data, shown in Figure 12.8.

A useful addition we have used here is the function `scale_fill_manual()` which supports specifying exact colors for the different categories. Here, the first color (where `Survived` is `FALSE`) is set to red, and the second color (`Survived` is `TRUE`) enables us to color the total number of survivors in green.

```
p6 <- ggplot(titanic,aes(x=Survived,fill=Survived))+
      geom_bar(color="white",alpha=0.75)+
      facet_grid(Sex~Class,scales="free")+
      theme(legend.position = "none")+
      labs(title="Survival outcomes on the Titanic",
```

```
            x="Survival Outcome",
            y="Number of Passengers")+
    scale_fill_manual(values=c("red","green"))
p6
```

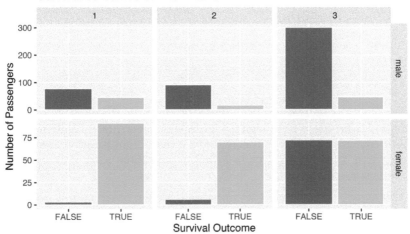

**FIGURE 12.8** Exploring the survival data from the Titanic dataset

This concludes our initial exploratory data analysis for the Titanic dataset. In a similar way to the iris dataset, because the suvival outcome is categorical, the data can be modelled using a decision tree classifier, which can generate rules that classify inputs (e.g., Sex, Age and Class) into outputs (either survived or perished) (Lantz, 2019). The `titanic` library supports this, as it contains both training and test datasets.

## 12.7 Exploring the effect of wind direction on winter temperatures

Our final example focuses on weather data, based on the tibble `observations` that is part of the `aimsir17` CRAN data package. Our question is to explore whether, in winter, wind from a southerly direction is more associated with higher temperatures. Before finalizing out dataset, we include three libraries required to support the analysis.

```
library(dplyr)
library(ggplot2)
library(aimsir17)
```

Given that there are 25 weather stations, to simplify our analysis we restrict the number to just four: Malin Head (north), Dublin Airport (east), Roches Point (south) and Mace Head (west).

The four character strings are stored in the vector st4.

```
st4 <- c("MALIN HEAD",
         "DUBLIN AIRPORT",
         "ROCHES POINT",
         "MACE HEAD")
st4
#> [1] "MALIN HEAD"     "DUBLIN AIRPORT" "ROCHES POINT"
#> [4] "MACE HEAD"
```

We then filter all observations from these stations from the dataset, and convert the units of wind speed from knots/hour to kilometers/hour. We do not include the year column, as this is redundant, given that all observations are from the calendar year 2017. The result is stored in the tibble eda0, and the first six observations are displayed.

```
# Filter data, convert to factor, and update average hourly wind speed
# from knots to kmh
eda0 <- observations %>%
        dplyr::filter(station %in% st4) %>%
        dplyr::mutate(station=factor(station),
                      wdsp=wdsp*1.852) %>%
        dplyr::select(-year)
head(eda0)
#> # A tibble: 6 x 11
#>   station month   day  hour date                  rain  temp  rhum
#>   <fct>   <dbl> <int> <int> <dttm>               <dbl> <dbl> <dbl>
#> 1 DUBLIN~     1     1     0 2017-01-01 00:00:00    0.9   5.3    91
#> 2 DUBLIN~     1     1     1 2017-01-01 01:00:00    0.2   4.9    95
#> 3 DUBLIN~     1     1     2 2017-01-01 02:00:00    0.1   5      92
#> 4 DUBLIN~     1     1     3 2017-01-01 03:00:00    0     4.2    90
#> 5 DUBLIN~     1     1     4 2017-01-01 04:00:00    0     3.6    88
#> 6 DUBLIN~     1     1     5 2017-01-01 05:00:00    0     2.8    89
#> # ... with 3 more variables: msl <dbl>, wdsp <dbl>, wddir <dbl>
```

Given that our initial question for exploration is to see whether there is an observable pattern between the wind direction and temperature, we now need to create a new feature to categorize the wind direction. To keep the analysis simple, we partition the compass space into four equal quadrants: north (N), east (E), south (S) and west (W), based on the logic shown below. For example, a wind direction greater than 45 degrees and less than or equal to 135 degrees is designated as east. We convert the wind direction to factors (N,E,S,W) in order to maintain a consistent appearance when the results are plotted (using a boxplot). The results are stored in the tibble eda.

```
eda <- eda0 %>%
    dplyr::mutate(wind_dir = case_when(
                    wddir > 315     | wddir <= 45    ~ "N",
                    wddir > 45      & wddir <= 135   ~ "E",
                    wddir > 135     & wddir <= 225   ~ "S",
                    wddir > 225     & wddir <= 315   ~ "W",
                    TRUE ~ "Missing"),
                wind_dir = ifelse(wind_dir=="Missing",
                    NA, wind_dir),
                wind_dir= factor(wind_dir,
                        levels=c("N","E","S","W"))) %>%
    dplyr::select(station:date,
            wdsp,
            wind_dir,
            wddir,
            everything())
dplyr::slice(eda,1:3)
#> # A tibble: 3 x 12
#>    station     month   day  hour date                       wdsp wind_~1
#>    <fct>       <dbl> <int> <int> <dttm>                     <dbl> <fct>
#> 1 DUBLIN AIR~     1     1     0 2017-01-01 00:00:00         22.2 N
#> 2 DUBLIN AIR~     1     1     1 2017-01-01 01:00:00         14.8 W
#> 3 DUBLIN AIR~     1     1     2 2017-01-01 02:00:00         14.8 W
#> # ... with 5 more variables: wddir <dbl>, rain <dbl>,
#> #    temp <dbl>, rhum <dbl>, msl <dbl>, and abbreviated variable
#> #    name 1: wind_dir
```

The features required for our analysis are now present, and therefore we can select the winter months and store the results in the tibble winter.

```
winter <- eda %>%
            dplyr::filter(month %in% c(11,12,1))
dplyr::glimpse(winter)
#> Rows: 8,832
#> Columns: 12
#> $ station   <fct> DUBLIN AIRPORT, DUBLIN AIRPORT, DUBLIN AIRPORT~
#> $ month     <dbl> 1, 1, 1, 1, 1, 1, 1, 1, 1, 1, 1, 1, 1, 1, 1, 1~
#> $ day       <int> 1, 1, 1, 1, 1, 1, 1, 1, 1, 1, 1, 1, 1, 1, 1, 1~
#> $ hour      <int> 0, 1, 2, 3, 4, 5, 6, 7, 8, 9, 10, 11, 12, 13, ~
#> $ date      <dttm> 2017-01-01 00:00:00, 2017-01-01 01:00:00, 201~
#> $ wdsp      <dbl> 22.22, 14.82, 14.82, 22.22, 20.37, 22.22, 24.0~
#> $ wind_dir  <fct> N, W, W, N, N, N, N, N, N, N, N, N, N, N, N, N~
#> $ wddir     <dbl> 340, 310, 310, 330, 330, 330, 330, 330, 330, 3~
```

```
#> $ rain       <dbl> 0.9, 0.2, 0.1, 0.0, 0.0, 0.0, 0.0, 0.0, 0.0, 0~
#> $ temp       <dbl> 5.3, 4.9, 5.0, 4.2, 3.6, 2.8, 1.7, 1.6, 2.0, 2~
#> $ rhum       <dbl> 91, 95, 92, 90, 88, 89, 91, 91, 89, 84, 84, 80~
#> $ msl        <dbl> 1020, 1020, 1020, 1020, 1020, 1020, 1020, 1021~
```

Our first analysis is to use `dplyr` to explore the median, 25th and 75th quantiles
for each wind direction and weather station. The code to perform this is listed
below, and again it highlights the power of `dplyr` functions to gather summary
statistics from the raw observations. The results are interesting in that the
wind directions from the south and west are in the first seven observations,
perhaps indicating that there is a noticeable temperature difference arising
from the different wind directions.

```
w_sum <- winter %>%
          dplyr::group_by(wind_dir,station) %>%
          dplyr::summarize(Temp25Q=quantile(temp,0.25),
                           Temp75Q=quantile(temp,0.75),
                           TempMed=median(temp)) %>%
          dplyr::arrange(desc(TempMed))
#> `summarize()` has grouped output by 'wind_dir'. You can override
#> using the `.groups` argument.
w_sum
#> # A tibble: 16 x 5
#> # Groups:   wind_dir [4]
#>    wind_dir station          Temp25Q Temp75Q TempMed
#>    <fct>    <fct>              <dbl>   <dbl>   <dbl>
#>  1 S        MACE HEAD           8.9    10.8     9.9
#>  2 S        ROCHES POINT        8.75   10.9     9.9
#>  3 W        MACE HEAD           7.3    10.1     8.6
#>  4 S        DUBLIN AIRPORT      6.32   10.2     8.3
#>  5 S        MALIN HEAD          5.5     9.6     8
#>  6 W        MALIN HEAD          6.4     9.2     8
#>  7 W        ROCHES POINT        5.3    10.6     7.9
#>  8 E        ROCHES POINT        6.3     9.1     7.5
#>  9 N        MALIN HEAD          5.35    7.6     6.6
#> 10 N        MACE HEAD           5       7.3     6
#> 11 E        MALIN HEAD          3.3     6.8     6
#> 12 E        MACE HEAD           3.48    7.8     5.6
#> 13 W        DUBLIN AIRPORT      2.5     8.4     5.5
#> 14 N        ROCHES POINT        3.8     8       5.4
#> 15 E        DUBLIN AIRPORT      2.4     7.3     5.35
#> 16 N        DUBLIN AIRPORT      1.4     6.05    4
```

We can build on this summary and present the temperature data using a box
plot, where we color and facet the data by wind direction. The code for this is
now shown, and the plot stored in the variable `p7`.

```
p7 <- ggplot(winter,aes(x=wind_dir,y=temp,color=station))+
        geom_boxplot()+
        facet_wrap(~station,nrow = 1)+
        labs(y="Temperature",
             x="Wind Direction",
             title="Winter temperatures at weather stations",
             subtitle="Data summarized by wind direction")+
        theme(legend.position = "none")
p7
```

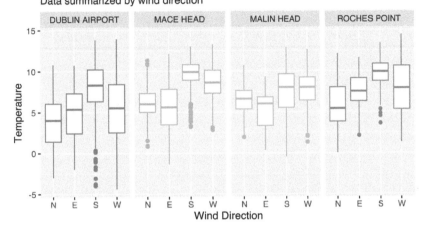

**FIGURE 12.9** Winter temperatures by station and wind direction

This plot, shown in Figure 12.9, highlights the power of the boxplot for comparing different results, as we can view the medians across all plots. A number of interesting observations arise:

- The coldest median wind directions for the four stations are north (Dublin Airport and Roches Point), and east (Mace Head and Malin Head).
- In all cases, the warmest median wind directions are from the south.
- The difference between the warmest and coldest, in all cases, is observable, indicating that the wind from the south seems associated with warmer weather conditions during the winter months from 2017.

Further statistical analysis would be interesting, and, similar to the Boston dataset, approaches such as ANOVA (Crawley, 2015) could be used to explore whether these differences are statistically significant.

## 12.8   Summary of R functions from Chapter 12

The following functions were introduced in this chapter.

| Function | Description |
| --- | --- |
| hour() | Retrieves the hour from a date-time object (lubridate). |
| month() | Retrieves the month from a date-time object (lubridate). |
| rename() | Changes columns name (dplyr). |
| stat_cor() | Adds correlation coefficients to a scatterplot (ggpubr). |
| wday() | Retrieves the weekday from a date-time object (lubridate). |
| yday() | Retrieves the day of year from a date-time object (lubridate). |
| year() | Retrieves the year from a date-time object (lubridate). |

# 13

## Linear Programming

The success of an OR technique is ultimately measured by the spread of its use as a decision making tool. Ever since its introduction in the late 1940s, linear programming (LP) has proven to be one of the most effective operations research tools.

— Hamdy A. Taha (Taha, 1992)

## 13.1 Introduction

Linear programming is a decision-making tool that has been ranked as among the most important scientific advances of the mid-20th century, and it is primarily focused on the general problem of allocating *limited resources* among *competing activities* (Hillier and Lieberman, 2019). Its success as a method arises from its flexibility in solving problems across many domains. It is also a foundational technique for other operations research methods, for example, integer programming and network flow problems. The technique uses a mathematical (linear) model to describe the problem of interest, and this model is then solved, using graphical (for simple cases), or computational techniques for larger, more realistic problems. Our goal in this chapter is to provide an introduction to using R to solve linear programming problems. Our initial focus is an example that can be explored in two dimensions, and solved graphically. We then show how this can be solved computationally using the R package lpSolve. We conclude the chapter with an introduction to sensitivity analysis for linear programming problems.

**Chapter structure**

- Section 13.2 provides a brief overview of the method, focusing on a standard way of representing problems.
- Section 13.3 presents the motivating example for this chapter, which is a linear programming model with just two decision variables (Taha, 1992).
- Section 13.4 demonstrates how R can be used to explore a subset of the feasible solution space to locate, and visualize, potential solutions.
- Section 13.5 presents a standard graphical solution to the two variable linear programming model.
- Section 13.6 shows how the two-variable problem can be solved using the R package `lpSolve`.
- Section 13.7 shows how sensitivity analysis can be used to explore the effect of parameter changes on the optimal solution.
- Section 13.8 provides a summary of the functions introduced in the chapter.

## 13.2  Linear programming: An overview

As mentioned in the introduction, linear programming is a technique that addresses the general problem of allocating *limited resources* among *competing activities*. For example, problems such as determining the number of customer service agents for a business to minimize costs, or finding the best use of raw material components to maximize profits can be addressed using the linear programming method. A linear programming problem is typically presented in a standard form (Hillier and Lieberman, 2019), which specifies the problem of allocating resources to activities, with the goal of selecting the values for the *decision variables* $x_1, x_2, ..., x_n$ so as to maximize $Z$, where:

$$Z = c_1 x_1 + c_2 x_2 + \cdots + c_n x_n$$

subject to the constraints

$$a_{11} x_1 + a_{11} x_2 + \cdots + a_{1n} x_n \leq b_1$$
$$a_{21} x_2 + a_{22} x_2 + \cdots + a_{2n} x_n \leq b_2$$

$$\vdots$$

$$a_{m1} x_2 + a_{m2} x_2 + \cdots + a_{mn} x_n \leq b_m$$
$$x_1 \geq 0, x_2 \geq 0, \cdots + x_n \geq 0$$
$$b_1 \geq 0, b_2 \geq 0, \cdots + b_m \geq 0$$

Note the following definitions of terms (Hillier and Lieberman, 2019):

- $Z$ is a value that contains the overall performance measure. The performance goal here can be maximization (e.g., profit), or minimization (e.g., cost). The function being maximized (or minimized) is also known as the *objective function*.
- $x_j$ is the level of activity $j$, where $(j = 1, \cdots, n)$.
- $c_j$ is the corresponding increase in $Z$ that results from a unit increase in the level of activity $j$.
- $b_i$ is the amount of resource $i$ that is available for allocation to activities, $(i = 1, \cdots, m)$.
- The number of activities is given by $n$, and the number of available resources to support the activities is given by $m$.
- The variables $x_1, \cdots, x_n$ are referred to as the **decision variables**, and the constants $c_j$,$b_i$ and $a_{ij}$ are known as the *model parameters*.

Furthermore, some additional terms are important when using the linear programming method. A *feasible solution* is one where all the constraints are satisfied, whereas an *infeasible solution* is one where at least one constraint is violated (Hillier and Lieberman, 2019). An *optimal solution* is therefore defined as a feasible solution with the maximum (for a maximization problem) or minimum (for a minimization problem) value of the objective function. Finally, the underlying modelling assumptions required for linear programming are important to keep in mind, and these are that the properties of *proportionality* and *additivity* are satisfied (Taha, 1992). Proportionality requires that the contribution of each decision variable to the objective function is directly proportional to the variable's value. Additivity requires that the objective function is the direct sum of the contributions of each decision variable.

## 13.3 The Reddy Mikks example

In order to demonstrate how we can use R to solve a linear programming problem, we will focus on the resource allocation problem known as the *Reddy Mikks* challenge (Taha, 1992), which has the following characteristics:

- The Reddy Mikks company is a hypothetical paint factory that produces two products, exterior and interior house paints, both targeted at wholesale distribution.
- Two core raw materials are required to produce the paints, and these are known as $A$ and $B$.
- Daily availability of the raw materials is limited, specifically: the maximum daily availability of $A$ is 6 tons, while $B$ has a maximum of 8 tons available for daily use.

- A ton of exterior paint requires 1 ton of $A$, and 2 tons of $B$.
- A ton of interior paint requires 2 tons of $A$, and 1 ton of $B$.

Further to these manufacturing details, additional information is available through a market survey:

- The daily demand for interior paint cannot be greater than demand for exterior paint by more than 1 ton.
- The maximum demand for interior paint is limited to 2 tons per day.
- The wholesale price for exterior paint is \$3000 per ton, while the price for interior paint is \$2000 per ton.

Given this information, the challenge is to find what amount of each paint to produce each day in order to *maximize gross income*. In order to find a solution, a number of questions should be addressed. First, we must identify the variables we want to find optimal solutions for, and in this example, those variables are to decide on the production amounts for (1) the quantity of exterior paint $x_1$, and (2) the quantity of interior paint $x_2$, where $x_1$ and $x_2$ are measured in units of tons/day. For linear programming, these are termed the *decision variables*.

Second, we identify the *objective function*, where we represent the total revenue as the variable $z$, and this is expressed as a linear combination of the decision variables $x_1$ and $x_2$. Given that the exterior paint generates revenues of \$3000 per ton, and interior paint generates \$2000 per ton, we can represent this revenue function as $Z = 3x_1 + 2x_2$ thousand dollars per day. The goal therefore is to determine values for $x_1$ and $x_2$ that maximize $Z$. We must ensure that the solution is feasible, and therefore does not violate any constraints described earlier. For example, given that we only have 6 tons of $A$ available per day, and that interior paint $x_2$ requires 2 tons per day of $A$, it would be infeasible to produce more that 3 tons of $x_2$ on any given day.

We now explore how all of the constraints can be represented algebraically.

- Raw material $A$ has a constraint in that its allocation between the two paints cannot exceed 6. Given that 1 unit of $A$ is needed for exterior paint, and two units are required for interior paint, this constraint can be represented as $x_1 + 2x_2 \leq 6$.

- Raw material $B$ has a constraint in that its total availability cannot exceed 8. We know that each ton of exterior paint requires 2 tons of $B$, whereas each ton of interior paint require 1 ton of $B$. This constraint can be represented as $2x_1 + x_2 \leq 8$.

- The first demand restriction is that the daily demand for interior paint cannot exceed that of exterior paint by more than 1 ton, and this can be expressed as $x_2 - x_1 \leq 1$.

- The second demand restriction is that the maximum demand for interior paint is 2 tons per day, and this can be expressed as $x_2 \leq 2$.

There are implicit constraints in that the production amounts for both paints cannot be negative, and therefore this can be expressed as $x_1 \geq 0$, and $x_2 > 0$

In summary, the full mathematical model for this optimization problem can be defined as follows:

To determine the tons of exterior and interior paint to be produced, expressed as the objective function:

$$Z = 3x_1 + 2x_2$$

Subject to the following constraints:

$$x_1 + 2x_2 \leq 6$$
$$2x_1 + x_2 \leq 8$$
$$x_2 + x_1 \leq 1$$
$$x_2 \leq 2$$
$$x_1 \geq 0$$
$$x_2 \geq 0$$

We will now focus on solving this problem, and the first step is to explore the range of feasible solutions using R.

## 13.4 Exploring a two-variable decision space using R

In order to provide an intuitive way to appreciate the linear programming process, and the search for an optimal solution, we will use our knowledge of R to explore permutations of the two decision variables in the Reddy Mikks problem, namely $x_1$ and $x_2$. In order to simplify the search process, we will constrain each of the two variables to be within the range $[0, 5]$, as we have prior information that an optimal solution will be in this area. Clearly, this would not be possible for more complex problems; however, it is viable for exploring this problem.

First, we include required libraries.

```
library(dplyr)
library(ggplot2)
```

We then define a function that determines whether a specific combination of values for $x_1$ and $x_2$ are feasible, given the four constraints. It performs a series of checks that return TRUE if the values satisfy all of the Reddy Mikks constraints.

```
is_feasible <- function (x1, x2){
  # Constraints for the Reddy Mikks Problem
  if((x1 + 2*x2 <= 6) &
     (2*x1 + x2 <= 8) &
     (-x1 + x2  <= 1) &
     (x2 <= 2)) TRUE
  else
    FALSE
}
# Check for x1=1, x2=2, should be TRUE
is_feasible(1,2)
#> [1] TRUE
# Check for x1=0, x2=2, should be FALSE
is_feasible(0,2)
#> [1] FALSE
```

In order to evaluate the profit from a particular solution, the objective function is needed. Here we code the function, and the linear combinations of $c_1 x_1$ and $c_2 x_2$ are returned.

```
# A function to evaluate the Reddy Mikks objective function
obj_func <- function(x1,x2){
  3*x1 + 2*x2
}
```

With these two functions created, we can now progress to explore the parameter space, and to see (1) which potential solutions are feasible, and (2) out of all the feasible solutions, which one provides the best result. Note, we do not call this the optimal result, as we are merely *exploring* the parameter space.

In this example, we create two vectors of possible values for $x_1$ and $x_2$, where we fix an arbitrary range of $[0, 5]$ for each, and we use the parameter length.out on the function seq() to generate two vectors of length 40, where each vector element is equidistant from its neighbor.

For example, here we create the vector x1_range. Note that its length is 40, starts at 0 and finishes at 5, and all its elements are equally spaced.

```
set.seed(100)
N = 40

x1_range   <- seq(0,5,length.out=N)
x1_range
#>  [1]  0.0000 0.1282 0.2564 0.3846 0.5128 0.6410 0.7692 0.8974
#>  [9]  1.0256 1.1538 1.2821 1.4103 1.5385 1.6667 1.7949 1.9231
#> [17]  2.0513 2.1795 2.3077 2.4359 2.5641 2.6923 2.8205 2.9487
#> [25]  3.0769 3.2051 3.3333 3.4615 3.5897 3.7179 3.8462 3.9744
#> [33]  4.1026 4.2308 4.3590 4.4872 4.6154 4.7436 4.8718 5.0000
```

We create a similar vector `x2_range`, and then we combine this with the vector `x1_range` using the R function `expand.grid()`, which creates a data frame containing all combinations of the two vectors ($40^2$ combinations). This output is copied into the tibble `exper`, where each row is a combination of parameter values.

```
x2_range   <- seq(0,5,length.out=N)
comb_x1_x2 <- expand.grid(x1_range,x2_range)
exper <- tibble(x1=comb_x1_x2[,1],
                x2=comb_x1_x2[,2])
str(exper)
#> tibble [1,600 x 2] (S3: tbl_df/tbl/data.frame)
#>  $ x1: num [1:1600] 0 0.128 0.256 0.385 0.513 ...
#>  $ x2: num [1:1600] 0 0 0 0 0 0 0 0 0 0 ...
```

Given that we now have our set of 1,600 combinations, we can add a new variable to the tibble `exp` that contains information on whether the solution is feasible or not. Because `is_feasible()` is not a vectorized function, we use the `dplyr` function `rowwise()` to ensure that the function `is_feasible()` is called for every row of data. The tibble's summary shows that 391 of the 1,600 possible solutions are feasible.

```
exper <- exper %>%
         dplyr::rowwise() %>%
         dplyr::mutate(Feasible=is_feasible(x1,x2))
summary(exper)
#>       x1             x2         Feasible
#>  Min.   :0.00   Min.   :0.00   Mode :logical
#>  1st Qu.:1.25   1st Qu.:1.25   FALSE:1209
#>  Median :2.50   Median :2.50   TRUE :391
#>  Mean   :2.50   Mean   :2.50
#>  3rd Qu.:3.75   3rd Qu.:3.75
#>  Max.   :5.00   Max.   :5.00
```

This information can be conveniently visualized using a scatterplot, shown in Figure 13.1, and each constraint line is drawn to clearly distinguish between feasible and non-feasible solutions.

```
p1 <- ggplot(exper,aes(x=x1,y=x2,color=Feasible))+
    geom_point(size=0.25)+
    geom_abline(mapping=aes(slope=-0.5,intercept=6/2),size=0.4)+
    geom_abline(mapping=aes(slope=-2,intercept=8),size=0.4)+
    geom_abline(mapping=aes(slope=1,intercept=1),size=0.4)+
    geom_abline(mapping=aes(slope=0,intercept=2),size=0.4)
p1
```

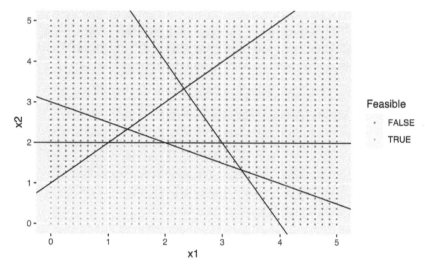

**FIGURE 13.1** Exploring the solution space for the Reddy Mikks problem

Our final step is to focus on the feasible solutions, and for each of these to calculate the profit via the objective function, and here we also show the top six solutions.

```
exper_feas <- filter(exper,Feasible==TRUE)

exper_feas <- exper_feas %>%
              dplyr::rowwise() %>%
              dplyr::mutate(Z=obj_func(x1,x2)) %>%
              dplyr::ungroup()

exper_feas %>% dplyr::arrange(desc(Z)) %>% head()
#> # A tibble: 6 x 4
#>      x1    x2 Feasible      Z
#>   <dbl> <dbl> <lgl>     <dbl>
#> 1  3.33  1.28 TRUE       12.6
```

```
#> 2  3.46 1.03  TRUE      12.4
#> 3  3.59 0.769 TRUE      12.3
#> 4  3.33 1.15  TRUE      12.3
#> 5  3.72 0.513 TRUE      12.2
#> 6  3.46 0.897 TRUE      12.2
```

We can extract the best solution from all of these, and then display it on a graph (shown in Figure 13.2) with the complete list of feasible points (from our original sample of 1,600 points).

```
max_point <- arrange(exper_feas,desc(Z)) %>%
             slice(1)

p2 <- ggplot(exper_feas,aes(x=x1,y=x2,color=Z))+geom_point()+
  scale_color_gradient(low = "blue",high = "red")+
  geom_abline(mapping=aes(slope=-0.5,intercept=6/2),)+
  geom_abline(mapping=aes(slope=-2,intercept=8))+
  geom_abline(mapping=aes(slope=1,intercept=1))+
  geom_abline(mapping=aes(slope=0,intercept=2))+
  geom_point(data = max_point,aes(x=x1,y=x2),size=3,color="green")+
  scale_x_continuous(name="x1", limits=c(0,5)) +
  scale_y_continuous(name="x2", limits=c(0,5))
p2
```

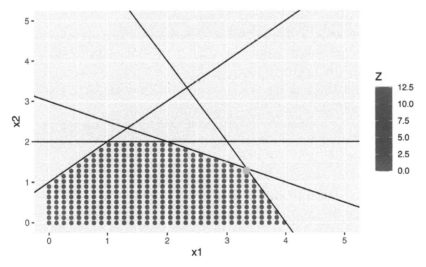

**FIGURE 13.2** Showing the feasible space and the best solution

Note that the best solution (from our sample) is close to the intersection of two of the lines on the top right of the feasible space. We will now see why this is significant, as we explore a graphical solution to this optimization problem.

## 13.5    Graphical solution to the Reddy Mikks problem

Graphical solutions to linear programming problems, while impractical for more complex problems, can provide important insights for smaller problems, especially those with just two decision variables. The initial step in constructing a graphical model is to plot the feasible space, and this is performed by graphing the line of each constraint (where the $\leq$ is replaced by $=$). This generates the graph displayed in Figure 13.3, a version of which we have already seen in the previous section, and this time the entire feasible space is shaded.

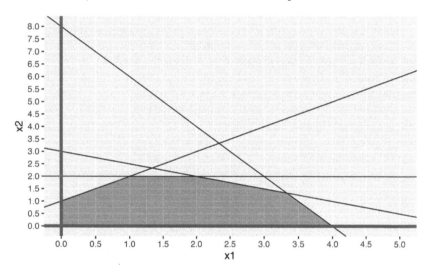

**FIGURE 13.3** Drawing the feasible space using the graphical method

In order to explore (graphically) which point in the feasible space is the optimal point, an interesting process is followed.

First, a new idea is presented, which is known as the *slope intercept* form of the objective function, where the objective function is re-formulated as the equation of a line of the form $x_2 = -c_1/c_2 \times x_1 + Z/c_2$.

Therefore, we can reformulate the Reddy Mikks objective function as: $x_2 = -(3/2) \times x_1 + Z/2$. The slope value is important, as the optimal solution must reside on a line with this slope, for a given value of $Z$. However, for the particular value of $Z$, the line must also intersect with a point (or points) in the feasible region.

We can now experiment with the following values of $Z$, and see what lines they generate. Note that all the lines constructed will be parallel, as they share the same slope of $-3/2$. The following table shows the options we have selected, and note that these are informed by a prior knowledge of the optimal solution.

| Selected Z Value | Intercept (Z/2) | Objective Function | Solutions |
|---|---|---|---|
| 14.666 | 7.333 | $x_2 = (-3/2)x_1 + 7.333$ | None |
| 12.666 | 6.333 | $x_2 = (-3/2)x_1 + 6.333$ | One |
| 10.666 | 5.333 | $x_2 = (-3/2)x_1 + 5.333$ | Infinite Number |
| 8.666 | 4.333 | $x_2 = (-3/2)x_1 + 4.333$ | Infinite Number |
| 6.666 | 3.333 | $x_2 = (-3/2)x_1 + 3.333$ | Infinite Number |
| 4.666 | 2.333 | $x_2 = (-3/2)x_1 + 2.333$ | Infinite Number |

These six lines are visualized on the graph, shown in Figure 13.4, with the optimal objective function line highlighted in green. This can be seen (by observation) to intersect the feasible region at the point $(3.333, 1.333)$, which is in fact the optimal solution. It is the optimal solution because it is the point on the objective function line that is within the feasible solution space, and it also yields the maximum value on the $x_2$ axis.

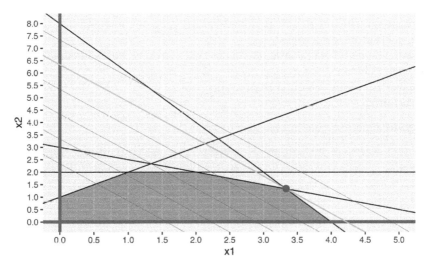

**FIGURE 13.4** Drawing parallel lines to locate the optimal solution.

The graphical solution process is a trial and error solution process, and once the slope is fixed, it involves finding the value in the feasible space that lies on this line, and also maximizes the intercept value. Once the intercept value is known, the maximum payoff (or profit) can be calculated, and in this case it is $6.3333 \times 2$ which evaluates to 12.667.

However, as pointed out, the graphical process is not practical for more than two decision variables. With more complex problems, there are two options. First, a manual process known as the *simplex method* can be used; however this process is outside of our scope in this chapter. Second, a computational solver can be utilized, and in R, the lpSolve package can be used to generate an optimal solution. We will now solve the Reddy Mikks example using lpSolve.

## 13.6   lpSolve: Generating optimal solutions in R

The first step is to include the linear programming library.

```
library(lpSolve)
```

In order to optimize a linear programming problem, the function lp() is called, and the required arguments are:

- direction, which is either "min" for minimization, or "max" for maximization, where the default value is "min".
- objective.in, which contains the numeric vector of coefficients for the objective functions (essentially the $c_i$ values).
- const.mat, which is a matrix of constraints, with one row per constraint, and one column for each variable.
- const.dir, a vector of character strings that provides the constraint directions, for example <=, <, >=, or >.
- const.rhs, which contains the number values for the right-hand side of the constraints.

Every decision variable is assumed to be $\geq 0$.

The lp() function returns an S3 object "lp", which contains the following elements.

| lp Element | Explanation |
|---|---|
| direction | Optimization direction, as specified in the call. |
| x.count | The number of decision variables in the objective function. |
| objective | The vector of objective function coefficients. |
| const.count | The total number of constraints provided. |
| constraints | The constraint matrix, as provided. |
| int.count | The number of integer variables (not used here). |
| int.vec | The vector of integer variable's indices (not used here). |
| objval | The objective value function at the optimum. |
| solution | The vector of optimal coefficients for the decision variables. |
| num.bin.solns | The number of solutions returned. |
| status | Return status, 0 = success, 2= no feasible solution. |

We now construct the linear programming formulation for the Reddy Mikks problem. First, we define the objective function in a vector. These are coefficient values for $c_1$ and $c_2$.

```
z <- c(3,2)
z
#> [1] 3 2
```

Next, we define the left-hand side of the four constraints. This is a $4 \times 2$ matrix, where each row represents the multipliers for each constraint, and each column represents a decision variable.

```
cons <- matrix(c(1,2,
                 2,1,
                 -1,1,
                 0,1),byrow = T,nrow=4)
cons
#>      [,1] [,2]
#> [1,]    1    2
#> [2,]    2    1
#> [3,]   -1    1
#> [4,]    0    1
```

Next, the direction of each constraint is specifed in a vector, where we have one entry for each constraint.

```
eql <- c("<=", "<=", "<=","<=")
eql
#> [1] "<=" "<=" "<=" "<="
```

Finally, the right-hand side values for the constaints are specifed, with four values provided, one for each constraint.

```
rhs <- c(6,8,1,2)
rhs
#> [1] 6 8 1 2
```

With these values specified, we can now call the function lp() to generate an optimal solution.

```
opt <- lp("max",z,cons,eql,rhs)
# Show the S3 class
class(opt)
#> [1] "lp"
# Show the status
opt$status
#> [1] 0
# Show the objective function value
opt$objval
```

```
#> [1] 12.67
# Show the solution points
opt$solution
#> [1] 3.333 1.333
```

As can be observed from the solution, it is the same as that calculated using the graphical method. It also confirms the utility of computational solutions to linear programming problems, as they are an efficient method for finding an optimal solution. However, one feature of the solution is that we assume that the objective function is fixed. With this in mind, it would be beneficial to see how the solution changes in response to changes in the objective function, and this process is known as sensitivity analysis.

## 13.7 Sensitivity analysis using lpSolve

Sensitivity analysis allows an analyst to explore the effect of parameter changes on the optimal solution (Taha, 1992). R is well-suited to providing the flexibility to conduct this type of analysis, and here we explore two types of sensitivity analysis and how they impact the range of optimal solutions:

- Varying the coefficients in the objective function, which will alter the slope of the objective function, and where it may intersect with the feasible solution space.
- Varying the right-hand side values of the constraints, which has the potential to either widen or narrow the feasible solution space.

These are now explored through the Reddy Mikks example.

### 13.7.1 Varying the objective function coefficients

In our first example, we vary the objective function coefficients, which essentially means changing the overall profit returned for exterior paint ($x_1$) and interior paint ($x_2$). To start, we include the relevant libraries.

```
library(lpSolve)
library(glue)
library(dplyr)
```

Next, we specify the linear programming problem, although we do not yet include the objective function.

```
cons <- matrix(c(1,2,
                 2,1,
                 -1,1,
```

```
                    0,1),byrow = T,nrow=4)

eql <- c("<=", "<=", "<=","<=")

rhs <- c(6,8,1,2)
```

We now define the range for the coefficient multipliers for the decision variables. These are arbitrary values, and we decide on four equidistant points between 1 and 10, and then create the full set of permutations, which yields 16 points.

```
c1_range    <- seq(1,10,length.out=4)
c2_range    <- seq(1,10,length.out=4)
comb_c1_c2 <- expand.grid(c1_range,c2_range)

of_exp <- tibble(x1_x=comb_c1_c2[,1],
                    x2_y=comb_c1_c2[,2])
of_exp
#> # A tibble: 16 x 2
#>      x1_x  x2_y
#>     <dbl> <dbl>
#>  1     1     1
#>  2     4     1
#>  3     7     1
#>  4    10     1
#>  5     1     4
#>  6     4     4
#>  7     7     4
#>  8    10     4
#>  9     1     7
#> 10     4     7
#> 11     7     7
#> 12    10     7
#> 13     1    10
#> 14     4    10
#> 15     7    10
#> 16    10    10
```

The optimal value for each of these combinations is then found by calling the function lp, where the second parameter, the objective function, is a combination of the profit multipliers for the two products. We also record the two solutions x1_opt and x2_opt, the payoff solution solution, the slope and intercept of each optimal solution, and a string version of the optimal point. Notice the value of dplyr in generating these solutions, as we can store the lp object as a list, and thereby store all the results in the tibble.

```
of_exp <- of_exp %>%
          dplyr::rowwise() %>%
          dplyr::mutate(lpSolve=list(lp("max",
                                        c(x1_x,x2_y),
                                        cons,
                                        eql,
                                        rhs)),
                        x1_opt=lpSolve$solution[1],
                        x2_opt=lpSolve$solution[2],
                        solution=lpSolve$objval,
                        slope=-x1_x/x2_y,
                        intercept=solution/x2_y,
                        opt_vals=glue("({round(x1_opt,2)},
                                      {round(x2_opt,2)})"))
```

The results can then be viewed, and here we arrange, from highest to lowest, by the solution value (the payoff.) We can observe that the highest payoff is obtained from the coefficients $(10, 10)$, with an optimal solution of $(3.333, 1.333)$.

```
of_exp %>% dplyr::arrange(desc(solution)) %>% head()
#> # A tibble: 6 x 9
#> # Rowwise:
#>     x1_x  x2_y lpSolve x1_opt x2_opt solution  slope intercept
#>    <dbl> <dbl> <list>   <dbl>  <dbl>    <dbl>  <dbl>     <dbl>
#> 1     10    10 <lp>      3.33   1.33     46.7  -1         4.67
#> 2     10     7 <lp>      3.33   1.33     42.7  -1.43      6.10
#> 3     10     1 <lp>      4      0        40   -10        40
#> 4     10     4 <lp>      4      0        40    -2.5      10
#> 5      7    10 <lp>      3.33   1.33     36.7  -0.7       3.67
#> 6      7     7 <lp>      3.33   1.33     32.7  -1         4.67
#> # ... with 1 more variable: opt_vals <glue>
```

An informative plot showing these optimal solutions is captured in Figure 13.5. It shows the shape of the optimal objective functions for the different variations, and it shows that three optimal solutions are found for the 16 different combinations of parameter values. It is interesting to see the location of optimal points on the extreme points of the graph, which is an expected output of the linear programming process.

## 13.7.2   Varying the right-hand side values for constraints

The second type of sensitivity analysis that we consider is to explore a range of parameters for the constraints, and in particular, exploring parameter changes for the constraint equations:

$$x_1 + 2x_2 \leq 6$$

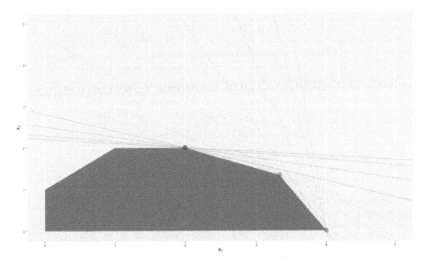

**FIGURE 13.5** Sensitivity analysis for the objective function

$x_2 \leq 2$

As a reminder, here is the linear programming problem, and in this case, the objective function will remain unchanged.

```
cons <- matrix(c(1,2,
                 2,1,
                 -1,1,
                 0,1),byrow = T,nrow=4)
eql <- c("<=", "<=", "<=","<=")
rhs <- c(6,8,1,2)
z <- c(3,2)
```

First, we explore changes in the first constraint, and vary the right-hand side value from its initial value of six, to an upper value of eight. Again, these choices are arbitrary. The dplyr code to run the sensitivity analysis is shown below, and in this case we vary the first element of the vector rhs. Note that the slope of the objective function is fixed at $-3/2$ because the objective function itself is fixed. The solutions are shown in the tibble cons1_exp, and the right-hand side (RHS) values of 7 and 8 yield the same optimal payoff of 13.

```
c1_range    <- seq(6,8,length.out=3)

cons1_exp <- tibble(c1_rhs=c1_range)

cons1_exp <- cons1_exp %>%
             dplyr::rowwise() %>%
```

```
              dplyr::mutate(lpSolve=list(lp("max",z,cons,eql,
                                 c(c1_rhs,rhs[2:4]))),
                      c1_rhs=c1_rhs,
                      x1_opt=lpSolve$solution[1],
                      x2_opt=lpSolve$solution[2],
                      solution=lpSolve$objval,
                      slope=-3/2,
                      intercept=solution/2,
                      opt_vals=glue("({round(x1_opt,2)},
                                    {round(x2_opt,2)})"))
cons1_exp
#> # A tibble: 3 x 8
#> # Rowwise:
#>   c1_rhs lpSolve x1_opt x2_opt solution slope intercept opt_vals
#>    <dbl> <list>   <dbl>  <dbl>    <dbl> <dbl>     <dbl> <glue>
#> 1      6 <lp>      3.33   1.33     12.7  -1.5      6.33 (3.33,
#> 1.~
#> 2      7 <lp>      3      2        13    -1.5      6.5  (3,
#> 2)
#> 3      8 <lp>      3      2        13    -1.5      6.5  (3,
#> 2)
```

We visualize these different solutions in Figure 13.6, which illustrates how the optimization process works, and specifically how changing the RHS can alter the optimal solution.

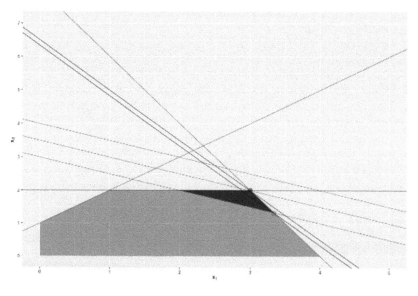

**FIGURE 13.6** Sensitivity analysis for the first constraint

A number of points are worth emphasizing:

- The two blue lines represent optimal solutions for the objective function. Note that the slope of these two lines is identical, which is what you would expect from an optimal solution where the objective function has not changed.

- The three red lines (slope of $0.5$) represent three different alternatives for the first constraint. The first red line (RHS = 6, intercept = 3) is the original constraint line that bounds the feasible space (shaded light gray), and the optimal solution here is the green dot $(3.33, 1.33)$.

- The second red line (intercept = 3.5) creates a new constraint line that also increases the feasible space, as highlighted by the darker shaded area. This new line intersects with the other constraint line $(x_2 \leq 2)$ at the point $(3, 2)$, and because of this, we have a new optimal point (the red dot), which is confirmed by the results returned from the function `lp()`.

- The third red line (RHS = 8, intercept = 4) does not change the feasible space in any way, as it intersects with $(x_2 \leq 2)$ at the point $(4, 2)$, and this point is well outside the feasible region. Therefore, the optimal solution at $(3, 2)$ holds for this new constraint. You can also observe that for the point $(4, 2)$ to be in the feasible region, the constraint shown by the line $2x_1 + x_2 = 8$ would have to move to the right by increasing its intercept value.

Finally, we explore changing the second constraint represented by the line $(x_2 = 2)$. Here we select two different values for the RHS, $(2, 1.0)$. The code to run the model is shown below.

```
c4_range   <- seq(1,2,length.out=2)
cons2_exp <- tibble(c4_rhs=c4_range)

cons2_exp <- cons2_exp %>%
              dplyr::rowwise() %>%
              dplyr::mutate(lpSolve=list(lp("max",
                                      z,
                                      cons,
                                      eql,
                                      c(rhs[1:3],c4_rhs))),
                        c4_rhs=c4_rhs, # Varying constraint 4
                        x1_opt=lpSolve$solution[1], # x1 solution
                        x2_opt=lpSolve$solution[2], # x2 solution
                        solution=lpSolve$objval, # Opt Z value
                        slope=-3/2, # Obj. func. slope
                        intercept=solution/2, # Obj. func. intercept
                        opt_vals=glue("({round(x1_opt,2)},
                                      {round(x2_opt,2)})"))
```

This generates the following optimal solutions.

```
cons2_exp
#> # A tibble: 2 x 8
#> # Rowwise:
#>    c4_rhs lpSolve x1_opt x2_opt solution slope intercept opt_vals
#>     <dbl> <list>    <dbl>  <dbl>    <dbl> <dbl>     <dbl> <glue>
#> 1       1 <lp>        3.5      1     12.5  -1.5      6.25 (3.5,
#> 1)
#> 2       2 <lp>       3.33   1.33     12.7  -1.5      6.33 (3.33,
#> 1.~
```

In a similar way to the last example, we can visualize these different solutions in Figure 13.7.

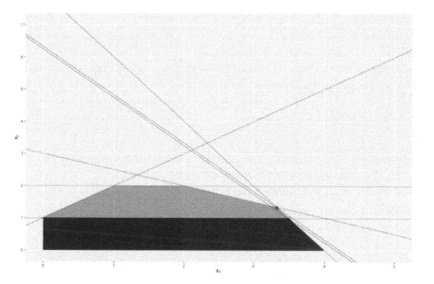

**FIGURE 13.7** Sensitivity analysis for the second constraint

The main points captured in this graph include:

- The blue lines represent the objective function at the two optimal solutions. Notice that the optimum values are very close, for example, the difference in the overall payoff is 0.17.

- The first (highest) red line represents the original constraint $x_2 \leq 2$, and we can see that the optimal solution is the red dot $(3.33, 1.33)$.

- The second (lowest) red line is the constraint $x_2 \leq 1.0$, and this leads to a new optimal solution represented by the green dot, which is the point $(3.5, 2)$. Notice that following the change in the constraint, the feasible region has also been altered, and this is highlighted by the darker shaded area.

To summarize, this section has shown how the optimal solution can move with changes in (1) the objective function coefficients and (2) the RHS values for the constraints. Software such as lpSolve also provides functionality to conduct forms of sensitivity analysis over more complex problem sets, and these approaches are also covered in more detail in (Taha, 1992) and (Hillier and Lieberman, 2010).

## 13.8   Summary of R functions from Chapter 13

The following R functions were introduced in this chapter.

| Function | Description |
|---|---|
| expand.grid() | Creates a data frame from combinations of vectors. |
| rowwise() | Processes a data frame one row at a time (dplyr). |
| lp() | R interface to the linear programming solver (lpSolve). |

# 14

## Agent-Based Simulation

> Situate an initial population of autonomous heterogeneous agents in a relevant spatial environment; allow them to interact according to simple local rules, and thereby generate - or "grow" - the macroscopic regularity from the bottom up.
>
> — Joshua M. Epstein (Epstein, 2012)

## 14.1 Introduction

Simulation is a valuable tool for modelling systems and an important method within the field of operations research. In this chapter we will focus on agent-based simulation, which involves the construction of a network of agents that can interact over time. Agent based simulation is often deployed to model diffusion processes, which are a common feature of many social, economic, and biological systems. In this chapter we build an agent-based model of product adoption, which is driven by a word of mouth effect when potential adopters interact with adopters. Upon completing the chapter, you should understand:

- The main elements of a graph, namely vertices and edges, and how a graph can be used to model a network structure.
- Four different types of network: fully connected, random, small world, and scale-free, and how to generate these networks using the igraph package.
- The agent design for the adopter marketing problem.
- The overall flow chart structure for the agent-based simulation.
- The R code for the model, including the network generation function and the simulation function.

- How to run many simulations using the function `furrr::map2_future()`, and then calculate overall simulation statistics such as quantiles for variables.
- Related R functions that allow you to create networks and run simulations.

### Chapter structure

- Section 14.2 provides an introduction to networks, and the R package `igraph`, with examples of four different network structures.
- Section 14.3 introduces the product adoption example, and the overall agent design. The adoption process is based on a diffusion mechanism originally used for infectious disease modelling, known as the Reed-Frost equation.
- Section 14.4 documents the simulator design, showing the overall flow chart and the key data structures used to record the simulation data.
- Section 14.5 presents the simulation code demonstrating both a single simulation run, and multiple runs.
- Section 14.6 provides a summary of all the functions introduced in the chapter.

## 14.2   Networks and the `igraph` package

A network is a collection of points joined together in pairs by lines, where the points are referred to as vertices (or nodes), and the lines are termed edges (Newman, 2018). Many real-world situations arising in business, biological, and social systems can be modelled as networks, for example; the spread of an infectious disease, the diffusion of a new product in the marketplace, and the interaction of manufacturers and distributors across a supply chain. Graph theory, described as the mathematical scaffold behind network science (Barabási, 2016), allows for the representation of a network as nodes and edges. The R package `igraph` facilitates this in a convenient way as it provides functions for both creating and analyzing graphs. Undirected graphs model a two-way relationship between two nodes, whereas a directed graph captures a one-way relationship. In our examples, we will model undirected graphs.

Before exploring several network structures, there are a number of network properties that are important to know:

- *Node degree*, which is the number of connections that an individual node has. In `igraph` the function `degree()` can be called to return the degree of each node in a network.

- *Average path length*, which is the average shortest distance between all pairs of nodes in the graph. This can vary depending on the network topology, and it can be calculated in `igraph` using the function `mean_distance()`.

To simplify our discussion of graphs, we will restrict our scope to four graph types, each with 20 nodes, and show how these can be generated using `igraph`. We execute the following code to set up our examples.

```
library(igraph)
N = 20
```

The four graphs are shown in Figure 14.1, and are now discussed, starting with the fully connected topology.

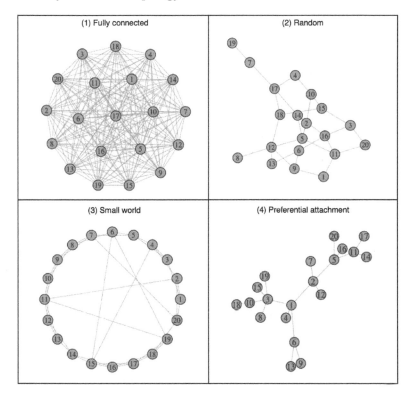

**FIGURE 14.1** Four network structures with N=20

- *Fully connected.* As the name suggests, a fully connected graph is one where each node is connected to all other nodes in the network. This network is visualized in part (1) of Figure 14.1, and shows how each node has the maximum number of connections. The `igraph` function `make_full_graph()` will create a fully connected graph. The results below confirm that each node in this network of 20 does indeed have 19 connections.

```
set.seed(100)
n1 <- make_full_graph(N)
degree(n1)
#>  [1] 19 19 19 19 19 19 19 19 19 19 19 19 19 19 19 19 19 19 19 19
```

```
mean_distance(n1)
#> [1] 1
```

- *Random.* A random graph, also known as the Erdős-Réyni model (Barabási, 2016), creates random connections between nodes. The network is visualized in part (2) of Figure 14.1, and we can observe that the degree profile of the random network nodes are significantly different from the fully connected network. For example, node 12 has five connections, while node 19 has just one connection. We use the function `sample_gnm()` to create a random graph. Here we conveniently specify the number of edges in the graph to be a multiple of the number of nodes (1.5 in this case). This is an arbitrary value to highlight the network structure. We will use a random graph structure for our agent-based simulation example.

```
set.seed(100)
n2 <- sample_gnm(n=N,m=N*1.5)
degree(n2)
#>  [1] 2 4 3 2 4 4 2 1 3 3 4 5 2 4 3 4 4 3 1 2
mean_distance(n2)
#> [1] 2.626
```

- *Small world.* This model generates a graph according to the Watts-Strogatz model (Watts and Strogatz, 1998). This approach takes a ring lattice structure with a neighborhood of two, and then randomly "rewires" connections. This property can be viewed in part (3) of Figure 14.1, where it can be seen that many nodes have four connections to their neighbors and their neighbor's neighbor. However, some of these links are "broken" and reset to another random node in the network. An interesting property of this graph is that it can be highly clustered yet still have a small path length, similar to random graphs. Because of this, the network can be deployed to modelling societal structures, where groups of people cluster into friendship networks, and some of the nodes have connections to other parts of the network. The function `sample_smallworld()` creates a small world network, and for this example, we provide a neighbor argument (`nei`) of two, and a rewire probability (`p`) of 0.10.

```
set.seed(100)
n3 <- sample_smallworld(1,N,nei=2,p=0.10)
degree(n3)
#>  [1] 3 4 3 4 4 4 4 4 4 4 6 4 3 4 5 4 4 4 4 4
mean_distance(n3)
#> [1] 2.242
```

- *Scale-free.* This model generates a network structure where new nodes are more likely to be connected to existing nodes with a higher degree (Barabási and Albert, 1999). This form of *preferential attachment* results in networks that share properties with real-world systems, for example, the number of connections of airports (i.e., the idea of a major hub in the transportation network), or a superspreader social contact network where one node (i.e., an infectious person) generates an unusually high amount of new infections. For this example, we generate the network with a call to `sample_pa()`, and specify that the network is not directed. The resulting network can be seen in part (4) of Figure 14.1, and this shows the preferential attachment taking hold, as node 3 has five connections, while many of the others only have a single connection.

```
set.seed(100)
n4 <- sample_pa(n=N,directed=F)
degree(n4)
#>  [1] 4 4 5 1 4 3 1 1 1 2 3 1 1 1 1 1 1 1 1 1
mean_distance(n4)
#> [1] 3.432
```

Networks play a key role in agent-based simulation, as they provide a topology for agents to interact and share information. We now focus on a design for an agent, based on the adopter marketing problem, where an agent's purchasing behavior is influenced by agents in their own network. As part of this example, we will utilize the random network structure; however, it would be straightforward to utilize an alternative network structure.

## 14.3 Agent design - The adopter marketing problem

The purpose of this agent-based simulation model is to explore how an individual's adoption decision is influenced by their neighbors. The example is inspired by the well-known Bass diffusion model (Bass, 2004), which is a differential equation-based model that simulates a product adoption process. Customers can be in one of two states, either potential adopters or adopters, and in our model we capture this structure from an agent-based perspective. Agent-based simulation is a stochastic process, which means there is an element of chance that determines whether an agent will adopt and transition from being a *potential adopter* to an *adopter*. In this model we assume that being an adopter is a final state, and once an agent reaches that state, they remain there.

Figure 14.2 provides an overall schematic of how this transition process works. Agent 7 is situated in a network with three contacts, which are agents 432, 898, and 974. Two of these agents are already adopters (898 and 974). The rounded dashed rectangle represents the agent's state. It is a *potential adopter* and will have a chance to adopt, based on the probability $\lambda$. The logic for switching is simple: we generate a random uniform number $U(0, 1)$ and if this value is less than $\lambda$, then a transition will happen. The challenge is to find a way to calculate $\lambda$, and to do this we use an equation from the field of infectious disease modelling known as the Reed-Frost equation. With this equation, the more adopters that are in an agent's immediate network, the higher the probability that the agent will adopt.

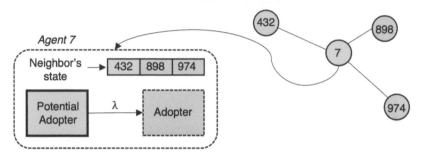

**FIGURE 14.2** The state transition mechanism for agent 7

The Reed-Frost equation has a number of elements (Vynnycky and White, 2010):

- First, there is the probability of adopting if a *potential adopter* makes contact with an *adopter*. Let's call this value $p$.

- Next, in order to adopt, an individual must be persuaded by at least one potential adopter. To formulate this, we consider the opposite case: the probability that a potential adopter is not persuaded by an adopter, which is $(1 - p)$. Working the logic from this, the probability that a potential adopter is not convinced to adopt by two adopters is $(1 - p)(1 - p)$. This leads to the general case that a potential adopter is not persuaded to adopt by all adopters as $(1 - p)^A$, where $A$ is the total number of adopters in their contact network.

- Finally, we can express the probability of adopting, $\lambda$ as the following time-varying value:

$$\lambda_t = 1 - (1 - p)^A$$

To illustrate how this probability works, consider the example in Figure 14.2, where we will calculate the probability of agent 7 adopting. Assume that the probability of being persuaded ($p$) is fixed at 0.10, and we know that agent 7 has two adopters in their network. Therefore, $\lambda$ takes on the value $1 - (1 - 0.1)^2$ which evaluates to 0.19.

Given that agent-based simulation is a stochastic process, we now draw a uniform random number, and if it is less than 0.19, agent 7 is persuaded by their contacts, and becomes an adopter. If the random draw is greater than 0.19, no change in state happens. This process is repeated for each time step in a simulation. However, in order to implement this process, we need to define (1) the overall logic of the simulation process and (2) a set of data structures to represent both the network and each agent's state.

## 14.4 Simulator design and data structures

To implement an agent-based simulation, we need to create a looping structure that runs for each simulation day and within that computes the chances of agents adopting, and we need to store information that can then be summarized. The simulation logic is shown in Figure 14.3. It involves an initialization stage where the contact network is created, and all agents initialized (for example, allocated one of two possible states: potential adopters or adopters).

Following that we have our *time loop*, and for each day in the simulation we:

- Get the latest list of potential adopters. We do this because we are only interested in the transition from potential adopter to adopter. So we will simulate each potential adopter interacting within their social network.
- For each potential adopter, use the Reed-Frost equation to calculate $\lambda$, the chance of adopting.
- Draw a random $U(0, 1)$ number, and if it is less than $\lambda$, mark the agent for change.
- At the end of the simulated day, convert all marked agents to adopters, and then increment the time by one.
- When the days are finished, organize the results and terminate the simulation.

For the simulation to work, we need to create R data structures to capture (1) the social network and (2) each agent's state and how this changes over time. This is an important design choice, and while there are many options available, our design decision is to use lists and tibbles. Tibbles are particularly useful for post-processing all the simulation data.

These data structures are shown in Figure 14.4. In part (a) we capture the full social network as a list of tibbles, where each list element contains an individual agent's connections. We use this approach for indexing convenience, as the name of each list element is the agent identifier, and the contents of the list show the agent's connections, in a tibble.

In part (b), we show three tibbles that will contain all of the generated simulation data. These are:

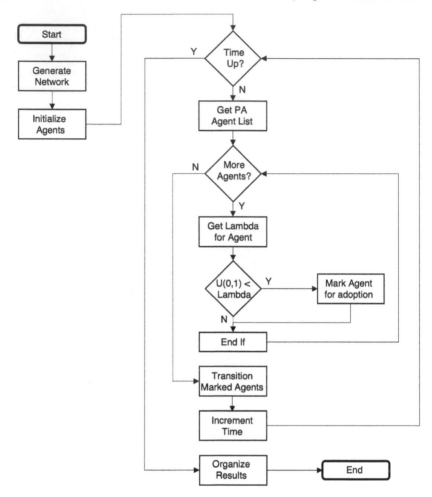

**FIGURE 14.3** Flow chart of the simulation algorithm

- `trace_sim` which contains agent information for each simulation, including the `agent_id`, the state logical flags `pa_state` and `a_state`, where only one of these will be `TRUE`, an attribute `change_to_a` used when an agent is marked for change, and `change_state_time` which records the simulation time when an agent changed state.

- `transitions` which is a record of when a state change occurred (i.e., when an agent transitioned from potential adopter to adopter) during a simulation run.

- `agent_sim_full`, where the tibbles `trace_sim` and `transitions` are joined to generate the complete simulation dataset.

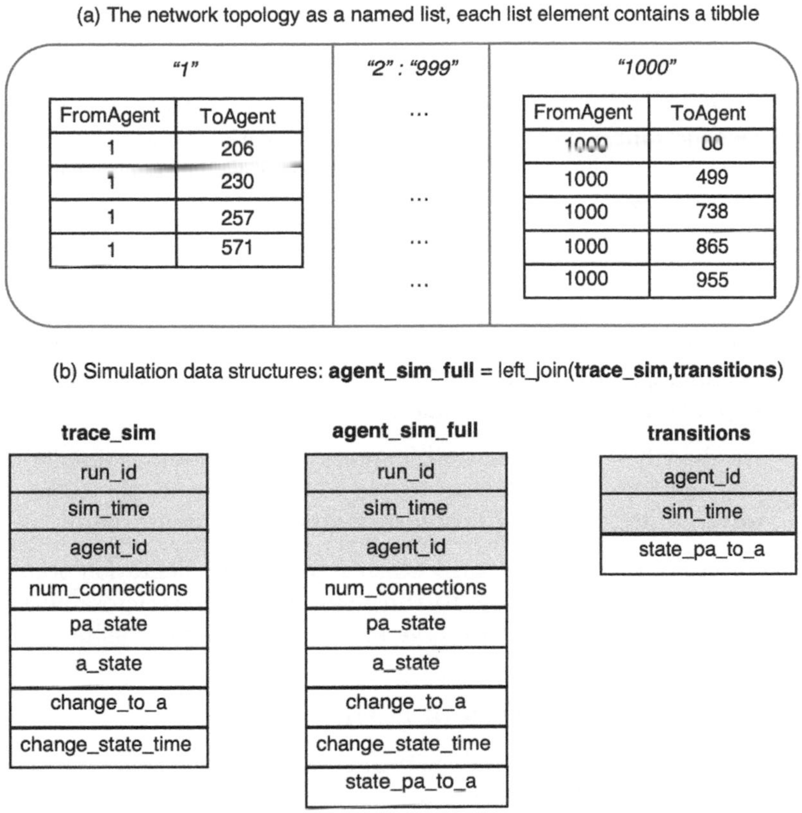

(a) The network topology as a named list, each list element contains a tibble

| "1" | | "2" : "999" | "1000" | |
|---|---|---|---|---|
| **FromAgent** | **ToAgent** | ... | **FromAgent** | **ToAgent** |
| 1 | 206 | | 1000 | 00 |
| 1 | 230 | | 1000 | 499 |
| 1 | 257 | ... | 1000 | 738 |
| 1 | 571 | ... | 1000 | 865 |
| | | ... | 1000 | 955 |

(b) Simulation data structures: **agent_sim_full** = left_join(**trace_sim**,**transitions**)

| **trace_sim** | **agent_sim_full** | **transitions** |
|---|---|---|
| run_id | run_id | agent_id |
| sim_time | sim_time | sim_time |
| agent_id | agent_id | state_pa_to_a |
| num_connections | num_connections | |
| pa_state | pa_state | |
| a_state | a_state | |
| change_to_a | change_to_a | |
| change_state_time | change_state_time | |
| | state_pa_to_a | |

**FIGURE 14.4** Data structure to support the simulation

Based on this design, we now explore the simulation code for the agent-based simulation.

## 14.5 Simulation code

For the simulation code, the following libraries are loaded:

```
library(dplyr)
library(purrr)
library(igraph)
library(ggplot2)
library(tidyr)
library(furrr)
```

The simulation code will be explained in four parts, specifically:

1. We look at how the network of agents is created. While using `igraph` there are many options for creating different network structures, we constrain our choice to a random network. We generate a list that contains the `igraph` object, and a list of agents along with their connections.

2. We explore the simulator *engine*, which is a function that iterates through time and can transitions agents from being potential adopters to adopters. This function can be called either once or many times. In all cases, this simulation function returns a tibble containing all the results, including each agent's state at every time step, and also the specific time of a transition event.

3. We show how the simulation function is called for one simulation run, and how the results can be processed to generate the adoption patterns, and also show how both states change over time.

4. We utilize the `furrr` package which supports parallelization of map functions. This allows us to run many simulations across all the computer's cores, and then organize and analyze the results.

### 14.5.1  Creating a network of agents

To create the consumer social network, we call the function `gener-ate_random_network()` which takes in four arguments (with defaults), and generates a list with three elements:

- `N`, the number of agents in the network.
- `graph`, the `igraph` representation of the network, and can be used to generate graphs, or in calling the `degree()` function which calculates the number of connections for each agent.
- `network`, a list of length `N` that contains, for each agent, a lookup tibble that shows the agent's neighborhood.

The function is shown below. Note that the `igraph` function `get.edgelist()` converts a network structure into a matrix, and this is then transformed to a data frame. This data frame, `gr_df` is then used to create a network tibble by calling the function `create_network_df()`.

```
generate_random_network <- function(N=1000,
                                     edge_mult=2,
                                     seed=F,
                                     seedVal=100){
  if(seed) set.seed(seedVal)
  graph      <- sample_gnm(n=N,m=N*edge_mult)
  gr_df      <- data.frame(get.edgelist(graph))
```

```
net_df      <- create_network_df(gr_df)
sim_network <- network_to_list(N,net_df)
list(N=N,
     graph=graph,
     network=sim_network)
}
```

An important requirement for our network model in the simulation is that the links for each agent are easily accessible. A drawback of the function `get.edgelist()` is that a link is represented in one row, and that the reciprocal link is not recorded (for example, if the link (1,2) is stored, the link (2,1) is not).

Therefore, we write the function `create_network_df()` which effectively doubles the size of the input data frame, and so it represents the complete set of links. This is needed because when we build our agent model, each agent will need to contain a reference to all its neighbors.

```
create_network_df<- function(df){
  network_df1 <- tibble(FromAgent=as.integer(df[,1]),
                        ToAgent=as.integer(df[,2]))
  network_df2 <- tibble(FromAgent=as.integer(df[,2]),
                        ToAgent=as.integer(df[,1]))

  network_df <- dplyr::bind_rows(network_df1, network_df2) %>%
                dplyr::arrange(FromAgent,ToAgent)
  network_df
}
```

We then perform a final transformation of the networked tibble by converting it to a list. The rationale for this is to reduce the search space for a lookup operation on the tibble, as we can first access an agent's list of connections using list indexing, and then process the tibble information. Essentially, this function creates a list of *N* tibbles. The code makes use of list processing operations specificed in Chapter 3, and uses the function `group_split()` covered in Chapter 10.

```
network_to_list <- function(N,net){
  agent_net <- vector(mode="list",length=N)
  names(agent_net) <- as.character(1:N)
  contacts <- group_split(net,FromAgent)
  names(contacts) <- map_chr(contacts,
                      ~first(as.character(.x$FromAgent)))
  for(nam in names(contacts)){
    agent_net[[nam]] <- contacts[[nam]]
  }
```

```
  agent_net
}
```

We can now observe how a network is created, and how we can use the information to: (1) show a frequency table of the number of connections across all agents (with a minimum of zero and a maximum of eleven), and (2) the specific connections for agents 1 and 1000, as they are previously represented in Figure 14.4.

```
net <- generate_random_network(1000,seed=T)
table(degree(net$graph))
#>
#>   0   1   2   3   4   5   6   7   8   9  10  11
#>  15  77 132 215 203 150  90  65  33  11   5   4
net$network[["1"]]
#> # A tibble: 4 x 2
#>    FromAgent ToAgent
#>        <int>   <int>
#> 1          1     206
#> 2          1     230
#> 3          1     257
#> 4          1     571
net$network[["1000"]]
#> # A tibble: 5 x 2
#>    FromAgent ToAgent
#>        <int>   <int>
#> 1       1000      88
#> 2       1000     499
#> 3       1000     738
#> 4       1000     865
#> 5       1000     955
```

With the network structure completed, we can now move on to exploring the R code that performs the agent-based simulation.

## 14.5.2   The simulation function

The simulation function is named run_sim(). It runs a single simulation based on the following inputs:

- The social network (net).
- The run identifier (run_id).
- A vector of agents who are initially adopters (adopters).
- The simulation end time (end_time).
- The probability of adoption given an interaction between a potential adopter and an adopter (prob_spread).

- Seed arguments to support replication of results (`seed` and `seed_val`).

The function implements the algorithm specified in Figure 14.3. The overall steps are:

- Create the tibble `agents` to store each agent's state (with $N$ rows), and initialize this with default values, for example, the column `pa_state` is set to TRUE for all potential adopters.
- Update the tibble `agents` to store the number of connections for each agent, and set all adopter agent's state `a_state` to TRUE.
- Create the `transitions` table (initially empty), and then add a transition to record all state changes for the initial adopters. To do this, the function `dplyr::add_row()` is used, as this is a convenient way to add a new row to a tibble. Every time an agent's state changes during the simulation, that information is appended to the `transitions` tibble.
- Create the tibble `trace_sim` that records the tibble `agents` for each simulation time. Note that the function `dplyr::bind_rows()` is used to "grow" this tibble during the course of the simulation.
- Enter the time loop and process each agent with a `pa_state` equal to TRUE. We only focus on those agents that can "flip" to the adopter state, as that will be sufficient for the simulation. Inside the loop, at each time step, the Reed-Frost equation is used to determine the probability of transitioning, and a random number is generated to determine whether or not the transition will occur.
- In processing the states, for convenience and efficiency we use matrix style subsetting to index and change the agent's attributes. For example, the following lines mark all agents in the vector `a` so that they will be "flipped" at a later stage, once the daily processing is completed.

```
agents[agents$agent_id==a,"change_to_a"] <- TRUE
```

- At the end of the simulation, the tibble `agent_sim_full` is created through a left join of `trace_sim` and `transitions`, and this tibble is returned by the function.

The code is now presented, with comments to highlight the main tasks.

```
run_sim <- function(run_id=1,
                    net,
                    adopters=c(1),
                    end_time=50,
                    prob_spread=0.10,
                    seed=FALSE,
                    seed_val=100){
  if(seed) set.seed(seed_val)

  agent_ids <- 1:net$N
```

```r
# Create a tibble for agent information
# This is stored for each time step
agents <- tibble(
  run_id              = rep(run_id,length(agent_ids)),
  sim_time            = rep(0L,length(agent_ids)),
  agent_id            = agent_ids,
  num_connections     = rep(0L,length(agent_ids)),
  pa_state            = rep(FALSE,length(agent_ids)),
  a_state             = rep(FALSE,length(agent_ids)),
  change_to_a         = rep(FALSE,length(agent_ids)),
  change_state_time   = rep(NA,length(agent_ids))
)

# Initialize all agents state, including adopter(s)
agents[,"num_connections"] <- as.integer(degree(net$graph))

agents[agents$agent_id %in% adopters,"pa_state"]        <- FALSE
agents[agents$agent_id %in% adopters,"a_state"]         <- TRUE
agents[!(agents$agent_id %in% adopters),"pa_state"]     <- TRUE
agents[!(agents$agent_id %in% adopters),"a_state"]      <- FALSE
agents[agents$agent_id==adopters,"change_state_time"]   <- 0L

# Create transitions table to log state changes for agents
transitions <- tibble(
  agent_id      = integer(),
  sim_time      = integer(),
  state_pa_to_a = logical()
)

# Record state changes for initial adopters
transitions <- dplyr::add_row(transitions,
                              sim_time=0L,
                              agent_id=as.integer(adopters),
                              state_pa_to_a=TRUE)

time <- 1:end_time
agents$run_id <- run_id

# Initialize tibble that will contain all states over time
trace_sim <- agents

# Start the simulation loop
for(t in time){
  agents$sim_time <- as.integer(t)
```

```r
# Get the list of potential adopters that have connections
# Return value as a vector (for convenience)
pa_list <- agents[agents$pa_state==TRUE &
                  agents$num_connections > 0,
                  "agent id",
                  drop=T]

# If there are potential adopters, loop through these
if(length(pa_list) > 0){

  for(a in pa_list){
    # Find all neighbors of agent a
    neighbors <- net$network[as.character(a)][[1]][,"ToAgent",
                                                   drop=T]

    # Find adopter neighbors (get agent ids as a vector)
    neigh_adopters <- agents[agents$agent_id %in% neighbors &
                             agents$a_state ==TRUE,
                             "agent_id",
                             drop=T]

    # Calculate the probability of adoption using Reed-Frost
    lambda <- 1 - (1 - prob_spread) ^ length(neigh_adopters)

    # Simulate random draw to decide if transition happens
    rn <- runif(1)
    if(rn < lambda ){
      # Mark agent for state change
      agents[agents$agent_id==a,"change_to_a"] <- TRUE
      # Record state change time
      agents[agents$agent_id==a,"change_state_time"] <- t
    }
  }
}

# End of day
# Flip states for those that have changed from pa to a
targets <- agents[agents$change_to_a==TRUE,"agent_id",
                  drop=TRUE]
agents[agents$change_to_a==TRUE,"pa_state"]    <- FALSE
agents[agents$change_to_a==TRUE,"a_state"]     <- TRUE
agents[agents$change_to_a==TRUE,"change_to_a"] <- FALSE

# Log the flip event as a transition
transitions <- dplyr::add_row(transitions,
```

```
                                      sim_time=as.integer(t),
                                      agent_id=as.integer(targets),
                                      state_pa_to_a=TRUE)


    }


    # Append agents states to overall list
    trace_sim <- dplyr::bind_rows(trace_sim,
                                  agents)
  } #End of simulation


  # Organize and return all results, including transition information
  agent_sim_full <- dplyr::left_join(trace_sim,
                                     transitions,
                                     by=c("sim_time","agent_id"))
}
```

With the simulation function defined, we can now explore how to run a single simulation, and analyze the results.

### 14.5.3 A sample run

Our first step is to create a network of 1,000 agents. The returned list contains the number of agents, the `igraph` network object, and the network list, which is used within the simulation.

```
net <- generate_random_network(1000,seed=T)
names(net)
#> [1] "N"         "graph"    "network"
table(degree(net$graph))
#>
#>   0   1   2   3   4   5   6   7   8   9  10  11
#>  15  77 132 215 203 150  90  65  33  11   5   4
```

Next, we run the simulation with a call to `run_sim()`, and we can observe the structure of the output, which is stored in `res`. There are 81,000 records, one for each agent at each time step, including their initial states.

```
res <- run_sim(net = net,
               end_time = 80)
dplyr::glimpse(res)
#> Rows: 81,000
#> Columns: 9
#> $ run_id         <dbl> 1, 1, 1, 1, 1, 1, 1, 1, 1, 1, 1, 1, 1~
#> $ sim_time       <int> 0, 0, 0, 0, 0, 0, 0, 0, 0, 0, 0, 0, 0~
#> $ agent_id       <int> 1, 2, 3, 4, 5, 6, 7, 8, 9, 10, 11, 12~
```

```
#> $ num_connections   <int> 4, 4, 5, 4, 7, 2, 3, 4, 0, 4, 3, 5, 4~
#> $ pa_state          <lgl> FALSE, TRUE, TRUE, TRUE, TRUE, TRUE, ~
#> $ a_state           <lgl> TRUE, FALSE, FALSE, FALSE, FALSE, FAL~
#> $ change_to_a       <lgl> FALSE, FALSE, FALSE, FALSE, FALSE, FA~
#> $ change_state_time <int> 0, NA, NA, NA, NA, NA, NA, NA, NA, NA~
#> $ state_pa_to_a     <lgl> TRUE, NA, NA, NA, NA, NA, NA, NA, NA,~
```

An important piece of information from the output is to record when an agent changed state, and the first ten agents can be viewed as follows.

```
res %>% dplyr::filter(agent_id %in% 1:10,
                      state_pa_to_a==TRUE) %>%
    dplyr::select(run_id,
                  agent_id,
                  sim_time,
                  state_pa_to_a)
#> # A tibble: 9 x 4
#>    run_id agent_id sim_time state_pa_to_a
#>     <dbl>    <int>    <int> <lgl>
#> 1       1        1        0 TRUE
#> 2       1        7       15 TRUE
#> 3       1        8       18 TRUE
#> 4       1       10       20 TRUE
#> 5       1        5       21 TRUE
#> 6       1        3       26 TRUE
#> 7       1        2       27 TRUE
#> 8       1        4       29 TRUE
#> 9       1        6       32 TRUE
```

This transition data can then be aggregated by day, to provide the total number of changes per day.

```
ar <- res %>%
    dplyr::group_by(sim_time) %>%
    dplyr::summarize(Adoptions=sum(state_pa_to_a,
                                   na.rm = T)) %>%
    dplyr::ungroup()
ar
#> # A tibble: 81 x 2
#>    sim_time Adoptions
#>       <int>     <int>
#> 1         0         1
#> 2         1         1
#> 3         2         1
#> 4         3         1
#> 5         4         1
```

```
#>  6          5              1
#>  7          6              2
#>  8          7              2
#>  9          8              4
#> 10          9              3
#> # ... with 71 more rows
```

We can then plot the adoption process, shown in Figure 14.5, to observe the number of transitions per day. This provides a summary of the "outbreak" which shows how the adoption has spread through the network over time.

```
p1 <- ggplot(ar,aes(x=sim_time,y=Adoptions))+geom_point()+geom_line()
p1
```

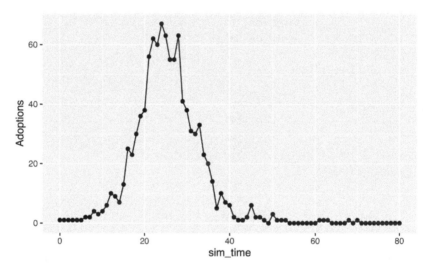

**FIGURE 14.5** Number of adoptions per day for one single run

The results data can also be used to visualize how the two states change over time. The following code sums both states for each time step, and then converts this to tidy data format. The result can be conveniently shown using geom_area(), and Figure 14.6 shows the diffusion graph for this simulation, and indicates that after 80 time units, most of the population are adopters (although not all).

```
states <- res %>%
           dplyr::group_by(sim_time) %>%
           dplyr::summarize(PA=sum(pa_state),
                            A=sum(a_state)) %>%
           tidyr::pivot_longer(PA:A,
                               names_to = "State",
                               values_to = "Number")
```

```
p2 <- ggplot(states,aes(x=sim_time,y=Number,color=State,fill=State))+
    geom_area()
p2
```

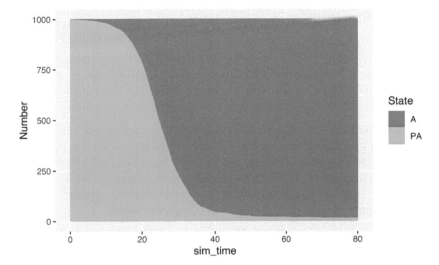

**FIGURE 14.6** Total number of potential adopters and adopters over time

For this particular simulation run, 17 agents did not adopt, and we can explore the network properties of those agents who have not adopted. When we take the agent state from the final time step (sim_time equals 80), it is interesting to see that 15 of the 17 non-adopting agents have zero connections, and the remaining two have just one connection. This confirms the expected behavior as potential adopter agents with zero connections cannot adopt, and agents with just one connection are less likely to adopt.

```
not_adopted <- res %>%
            dplyr::filter(sim_time==80,pa_state==TRUE) %>%
            dplyr::pull(num_connections) %>%
            table()
not_adopted
#> .
#> 0
#> 15
```

In summary, the simulation run is configurable, and returns the results in a tibble, which can then be analyzed. However, as a guide, the results of a single stochastic simulation provides only one possible trajectory. To increase confidence in the results from agent-based simulations, it is recommended to run the model many times, and to explore the overall statistics over multiple runs. We now explore this option of running the model many times.

### 14.5.4   Analyzing multiple runs

We will now explore multiple runs with experiments designed based on three different initial conditions that relate to the seeding strategy for identifying the initial adopter agent, namely:

- Seed the network with an adopter agent with only one connection. We will label this experiment *Lowest Connections*.
- Seed the network with an adopter agent with the maximum number of connections, in an experiment labelled *Highest Connections*.
- Seed the network with an adopter agent that has the mean number of connections, where this experiment is called *Mean Connections*.

For each simulation the network structure remains unchanged, but the draws of random numbers that govern the adoption transitions will be different (as we do not fix the seed for each simulation), and the initial adopter will be one of the three alternatives. In order to identify three agents, we take a vector of agent degrees via the `igraph` object, and select the lowest, highest and mean, and these agents ids are then stored in the variable `inits`.

```
net <- generate_random_network(1000,seed=T)
conns <- degree(net$graph)
inits <- c(Lowest=which(conns == 1)[1],
           Highest=which(conns == max(conns))[1],
           Mean=which(conns == floor(mean(conns)))[1])
inits
#>  Lowest Highest    Mean
#>      20     251       1
```

As running the simulations is computationally intensive, we make use of the `furrr` package which offers the facility to run map functions across the computer's cores. To start this, the function `future::plan()` is called, and the argument `multisession` means that the `furrr` package will then distribute the simulations over all available cores.

```
library(furrr)
library(future)
plan(multisession)
```

Next, we run the set of simulations, in this example, 150 are run. The simulations sample from the trio of agents specified in `inits`, and the function `furrr::future_map2()` is used, which provides the exact same behavior as `purrr::map2()` except that it enables you to map in parallel across all available cores. Note that the argument `.options = furrr_options(seed=T)` needs to be added for random number generating purposes.

```
NSim <- 150
sims <- furrr::future_map2(1:NSim,
```

```
                         rep(inits,NSim/3),
                         ~run_sim(run_id    = .x,
                                  net       = net,
                                  adopters  = .y,
                                  end_time  = 80),
                         .options = furrr_options(seed = T)) %>%
          dplyr::bind_rows()
sims
#> # A tibble: 12,150,000 x 9
#>     run_id sim_t~1 agent~2 num_c~3 pa_st~4 a_state chang~5 chang~6
#>      <int>   <int>   <int>   <int> <lgl>   <lgl>   <lgl>     <int>
#> 1        1       1       0       1       4 TRUE    FALSE   FALSE      NA
#> 2        2       1       0       2       4 TRUE    FALSE   FALSE      NA
#> 3        3       1       0       3       5 TRUE    FALSE   FALSE      NA
#> 4        4       1       0       4       4 TRUE    FALSE   FALSE      NA
#> 5        5       1       0       5       7 TRUE    FALSE   FALSE      NA
#> 6        6       1       0       6       2 TRUE    FALSE   FALSE      NA
#> 7        7       1       0       7       3 TRUE    FALSE   FALSE      NA
#> 8        8       1       0       8       4 TRUE    FALSE   FALSE      NA
#> 9        9       1       0       9       0 TRUE    FALSE   FALSE      NA
#> 10      10       1       0      10       4 TRUE    FALSE   FALSE      NA
#> # ... with 12,149,990 more rows, 1 more variable:
#> #   state_pa_to_a <lgl>, and abbreviated variable names
#> #   1: sim_time, 2: agent_id, 3: num_connections, 4: pa_state,
#> #   5: change_to_a, 6: change_state_time
```

The returned list of tibbles is then collapsed into one large data frame using the function `dplyr::bind_rows()`. We can also observe that there are 12,150,000 rows in the tibble, which is as expected (1000 Agents × 81 time steps × 150 simulations). This tibble is then processed to generate results.

First, we can generate summary data for the adoptions across the three different sets of initial conditions, and for convenience, these are also labelled to clarify the outputs. Patterns are evident when exploring the plots, shown in Figure 14.7. Specifically, when the highest connected agent is seeded, the adoption process takes off faster and produces and earlier peak, and when the lowest connected agent is seeded, it results in a later peak and much more variability in adoption trajectory.

```
ar <- sims %>%
      dplyr::group_by(run_id,sim_time) %>%
      dplyr::summarize(Adoptions=sum(state_pa_to_a,
                                     na.rm = T)) %>%
      dplyr::mutate(run_desc=case_when(
                    run_id %% 3 == 1 ~ "Lowest Connections",
                    run_id %% 3 == 2 ~ "Highest Connections",
```

```
                              run_id %% 3 == 0 ~ "Mean Connections")) %>%
          dplyr::ungroup()


p1 <- ggplot(ar,aes(x=sim_time,y=Adoptions,
                    group=run_id,color=run_desc))+
          geom_point()+geom_line()+facet_wrap(~run_desc,nrow = 3)
p1
```

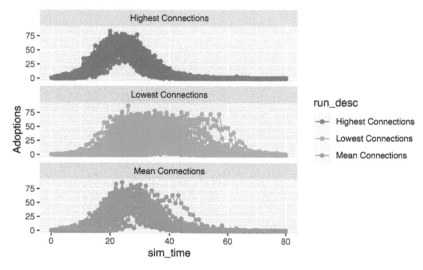

**FIGURE 14.7** Simulation trajectories for each run

A common approach used in agent-based simulation is to report the quantiles from the full collection of simulation outputs. This can support generation of insights. Here we find the 5% and 95% quantiles, and so include 90% of all the results, for all the runs across each simulation time, for each initial seeding scenario. These results are stored in the tibble `quants`.

```
quants <- ar %>%
              dplyr::group_by(run_desc,sim_time) %>%
              dplyr::summarize(Q05=quantile(Adoptions,0.05),
                               Q95=quantile(Adoptions,0.95),
                               Mean=mean(Adoptions)) %>%
              dplyr::ungroup()


quants
#> # A tibble: 243 x 5
#>    run_desc              sim_time   Q05   Q95  Mean
#>    <chr>                    <int> <dbl> <dbl> <dbl>
#> 1 Highest Connections          0     1     1     1
#> 2 Highest Connections          1     0     2     1
```

```
#>  3 Highest Connections    2 0      2.55  1.14
#>  4 Highest Connections    3 0      4     1.52
#>  5 Highest Connections    4 0      5     2.18
#>  6 Highest Connections    5 0      7     3.06
#>  7 Highest Connections    6 1      6     2.96
#>  8 Highest Connections    7 0.45   8     4.24
#>  9 Highest Connections    8 0.45   11    5.26
#> 10 Highest Connections    9 1      13.5  6.84
#> # ... with 233 more rows
```

The `quants` tibble can be visualized, and the 90% interval conveniently highlighted using the function `geom_ribbon()`. The visualization is shown in Figure 14.8.

```
p4 <- ggplot(quants,aes(x=sim_time,y=Mean,
                        color=run_desc,group=run_desc))+
        geom_ribbon(aes(ymin=Q05,
                        ymax=Q95,
                        fill=run_desc,
                        group=run_desc),
                     alpha=0.2)+
    geom_line(size=2)+
  theme(legend.position = "top")

p4
```

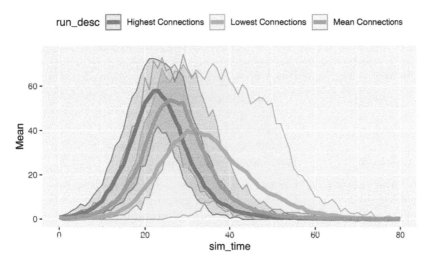

**FIGURE 14.8** Summary of quantiles for each run

The quantile analysis reveals a number of insights:

- The means of the three are different, with the simulations with the highest connected adopter leading to a simulation with a higher peak and an earlier peak time.
- The quantiles are narrower around the highest connections scenario, and get progressively wider with the remaining scenarios. This indicates a more predictable outcome when the agent with the highest number of connections is the initial adopter, which is intuitive, as such an agent could be viewed a form of "superspreader" in the network.

In summary, the ability to run many simulations is an integral part of the agent-based simulation approach. Efficient design of the simulation function, and alignment with the `furrr::map2()` function, yields a straightforward process whereby many simulations can be run, and the results merged into a large tibble for further analysis. While the model here is quite simple, it could be built upon to model an infectious disease outbreak, for example, a susceptible-infected-recovered (SIR) model. For more complex models, there are frameworks that can be used, for example, the open source agent-based simulation tool NetLogo (Wilensky and Rand, 2015), which has been used in domains such as epidemiology, and could also be used to implement our agent design presented in this chapter.

## 14.6  Summary of R functions from Chapter 14

A list of functions introduced in this chapter is now presented.

| Function | Description |
|---|---|
| future_map2() | A parallel implementation of purrr::map2() (furrr). |
| plan() | Used to specify how futures are resolved (future). |
| geom_ribbon() | Display y interval set by ymin and ymax (ggplot2). |
| add_row() | A function to add rows of data to a tibble (tibble). |
| bind_rows() | Binds many data frames into one (dplyr). |
| make_full_graph() | Creates a fully connected graph (igraph). |
| sample_gnm() | Creates a random graph (igraph). |
| sample_smallworld() | Creates a small-world graph (igraph). |
| sample_pa() | Creates a scale-free graph (igraph). |
| get.edgelist() | Returns a matrix of edges (igraph). |
| degree() | Returns the number of vertex edges (igraph). |

# 15

## System Dynamics

Everything we do as individuals, as an industry, or as a society
is done in the context of an information-feedback system.

— Jay W. Forrester (Forrester, 1961)

## 15.1   Introduction

In Chapter 14 we presented agent-based simulation, which focused on modelling
the decisions of individuals interacting withing a social network structure. We
now present a complementary simulation method known as system dynamics,
which is used to build models that can support policy analysis and decision
making. It has been successfully deployed across a range of application areas,
including health care, climate change, project management, and manufacturing.
The system dynamics approach is grounded in calculus (i.e., a model as a system
of ordinary differential equations) and takes the perspective of modelling a
system by focusing on its stocks, flows, and feedback. This chapter provides an
overview of system dynamics, and demonstrates how system dynamics models
can be implemented in R, using the deSolve package. For a comprehensive
perspective on the method, many valuable textbooks can be consulted, for
example (Sterman, 2000; Morecroft, 2015; Warren, 2008).

Upon completing the chapter, you should have an appreciation for:

- The main components of a system dynamics model: stocks, flows, and feedback.
- The classic limits to growth model, its stock and flow structure, and representation as a system of ordinary differential equations (ODEs).

- How to configure and run a system dynamics model using the `deSolve` package, and in particular, the function `ode()`.
- The Susceptible-Infected-Recovered (SIR) stock and flow model, and its underlying assumptions, including the transmission equation.
- An extension to the SIR model, the susceptible-infected-recovered-hospital (SIRH) model, which considers the downstream effects of a novel pathogen on a hospital system.
- The formulation of two policy countermeasures, and the use of sensitivity analysis to explore their interaction.
- Related R functions that support building system dynamics models in R.

### Chapter structure

- Section 15.2 provides an overview of system dynamics as a simulation method, and how this method is based on solving a system of ordinary differential equations (ODEs).
- Section 15.3 introduces the R package `deSolve`, which provides a set of general solvers for ODEs.
- Section 15.4 presents the Susceptible-Infected-Recovered (SIR) model of infectious disease transmission, and shows how this can be implemented using the `deSolve` function `ode()`.
- Section 15.5 extends the SIR model to include a hospital stream, so that downstream effects such as hospital admissions can be modelled. It also adds two countermeasures to the model: one medical (vaccination), and the other non-pharmaceutical (mobility reduction).
- Section 15.6 shows how sensitivity analysis can be performed with two parameters: one that influences population mobility, and the other that impacts the speed of vaccination. Using the tools of the `tidyverse` the results are presented that show the overall impact of countermeasures, based on measuring the maximum peak of patients in hospital.
- Section 15.7 provides a summary of all the functions introduced in the chapter.

## 15.2   Stocks, flows, and feedback

### 15.2.1   Stocks and flows

The stock is a central idea in system dynamics, and is defined as a component of a system that accumulates or drains over time (Ford, 2019). Examples of stocks include:

- The balance of a bank account, which accumulates lodgements and interest added, and it drains when withdrawals are made.

- The number of employees in a firm, which accumulates when new people are hired, and reduces when people leave.
- The amount of inventory in a warehouse, which accumulates when new inventory arrives, and drains when items are sent to customers. We can measure inventory using the term stock keeping unit (sku).
- The number of people infectious with a virus, which accumulates when people get infected and is reduced when people recover and are no longer infectious.
- The amount of water in a reservoir, which accumulates with rainfall, and drains with evaporation, or when the valves are opened to release water.

An observation from these five examples is that they relate to many different domains, for example, business, health, and the environment. In other words, *stocks exist in many different fields of study*, and therefore we can model different systems using the system dynamics method. The example of a reservoir provides a metaphor for stock and flow systems, often expressed with the related idea of a bathtub containing water (i.e., the stock), and that the level of water can rise through water flowing in (via the tap), and fall when water leaves (via the drain). Another point worth noting is that when building models in system dynamics, we also record the *units* of a variable, for example, the units for the stock employees are *people*. These units are important, as they must balance with the units of a flow, which we now define.

The bathtub metaphor also encourages us to think about how stocks change, and for a bathtub, the water level changes through inflows and outflows. For example, if the drain remains fully sealed and there is no flow from the taps, and there is no other way for the water to dissipate, then the water level will remain unchanged. If, however, the drain is open, and no water flows in to the bathtub, then the water level will fall. The main idea here is that the level of water can only change through either the tap or the drain (we are assuming that there is no evaporation or people removing water through another container).

The tap and the drain have there equivalents in a model: they are called flows. A flow is defined as the movement of quantities between stocks (or across a model boundary), and therefore flows represent activity (Ford, 2019). For each of the five examples we just introduced, the flows that change the stock can be identified, and are shown below. Note that the units of change for a flow are the units of the relevant stock *per time period*.

| Stock | Inflow(s) | Outflow(s) |
|---|---|---|
| Bank account (euro) | Lodgements (euro/day) | Withdrawals (euro/day) |
| Employees (people) | New hires (people/year) | Retirements, resignations (people/year) |
| Inventory (sku) | Incoming shipments (sku/week) | Outgoing shipments (sku/week) |

| Stock | Inflow(s) | Outflow(s) |
|---|---|---|
| People infected (people) | People getting infected (people/day) | People recovering (people/day) |
| Reservoir (litres) | Rainfall(litres/minute) | Evaporation (litres/minute) |

We can summarize three principles of stock and flow systems:

- If the *sum of all inflows* exceeds the *sum of all outflows*, the stock will rise.
- If the *sum of all outflows* exceeds the *sum of all inflows*, the stock will fall.
- If the *sum of all outflows* equals the *sum of all inflows*, the stock level will remain constant, and this condition is known as *dynamic equilibrium*.

Stocks and flows have their direct equivalent in calculus, where the stock is the integral, and the *net flow* is the derivative. Therefore, if the model can be integrated using the rules of calculus, we can generate results this way. For example, the net flow of a model to predict the balance of an interest-bearing account over time, with no withdrawals, can be defined as $dB/dt = rB$, where $B$ is the balance, and $r$ the interest rate. The solution to this, using the rules of integration (and outside this chapter's scope), is $B_t = B_0 e^{rt}$, where $B_0$ is the initial account balance. However, most system dynamics models rely on a numerical integration algorithm to simulate the values, and one method for doing this is to use Euler's equation. In this chapter we do not implement this algorithm; instead we invoke an implementation of it using R's deSolve package, via the function ode().

### 15.2.2   Simulating stocks and flows

To provide an intuition about how numerical integration works, we explore Euler's method for generating simulation results. The method can be described by the equation $S_t = S_{t-dt} + NF * dt$, where $S$ is the stock we are modelling, $dt$ is the time step (usually around 0.25 or less), and $NF$ is the net flow for the stock. This equation models the stock's accumulation process, and we can show how this works by taking the simple bank account interest example. We have built a simple model in deSolve and here are the results (note that we describe how to build these models in Section 15.3).

```
res
#>   time     B    r Net_Flow  DT
#> 1  0.0 100.0 0.1    10.00 0.5
#> 2  0.5 105.0 0.1    10.50 0.5
#> 3  1.0 110.2 0.1    11.03 0.5
#> 4  1.5 115.8 0.1    11.58 0.5
#> 5  2.0 121.6 0.1    12.16 0.5
```

```
#> 6  2.5 127.6 0.1    12.76 0.5
#> 7  3.0 134.0 0.1    13.40 0.5
```

The results are shown for the first 3 years, and the value for $dt$ is 0.5. The initial account balance is 100, and the interest rate is 0.1. We can work through the solution from left to right and explore the first two rows:

- At time = 0.0, the stock value is 100.00. This is often termed the *initial condition* of the stock, and for integration to work, initial conditions for all the stocks must always be specified.
- The net flow for the duration $[0, 0.5]$ is then calculated, and this is simply the right-hand side of the net flow equation $rB$, and this evaluates to 10.0.
- The next calculation is for the stock at time 0.5. Here we use Euler's equation so that $S_{0.5} = S_0 + 10 * 0.5$, and this evaluates to 105. If we can imagine this on a graph, we would see that we are adding the area of a rectangle to the stock, where the height is 10.0 and the width 0.5.

Therefore, when we simulate using system dynamics, we are invoking a numerical integration algorithm to solve the system of stock and flow equations. Other integration approaches can be used, for example, Runge-Kutta methods, which are also available in the R package deSolve.

### 15.2.3 Feedback

Before considering our first model, and its simulation using deSolve, we briefly summarize an additional concept in system dynamics, which is *feedback* (Sterman, 2000; Richardson, 1991). Feedback occurs in a model when the effect of a causal impact comes back to influence the original cause of that effect (Ford, 2019). For example, the net flow equation $dB/dt = rB$ contains feedback, because the net flow depends on the stock, and the stock, in turn, is calculated from the net flow. We can describe these relationships as follows:

- An increase in the stock $B$ leads to an increase in the flow $dB/dt$. This is an example of *positive polarity*, where the variables move in the same direction. These relationships can be shown on a stock and flow diagram. (If the cause and effect relationship was negative, for instance if the variables moved in opposite directions, then it would be termed negative polarity.)

- An increase in the flow $dB/dt$ causes an increase in the stock $B$. That is because a stock can only change through its flows (a rule of calculus).

These two relationships give rise to a *feedback loop*; in this case, it is a *positive feedback* loop, because the variable $P$ is amplified following an iteration through the loop. Positive feedback loops, if left uncorrected, give rise to exponential growth behavior. In system dynamics, there is an additional type of loop known as a *negative feedback* (or balancing) loop, where a variable would move in the opposite direction following an iteration through a loop.

Therefore, the polarity of a loop can be either positive or negative, and it can be found by taking the algebraic product of all signs around a loop (Ford, 2019). An advantage of stock and flow diagrams is that feedback loops can be visualized, and the following limits to growth example shows both a positive and negative feedback loop in the stock and flow system.

## 15.2.4   Limits to growth model

Before moving on to running simulation models using deSolve, we present a specification of a well-known stock and flow model, which is a model that explores limits to growth for a stock. This model is most widely known as the Verhulst model of population growth, published by the Belgian mathematician Pierre Verhulst in the 1830s. The stock and flow representation of this model is shown in part (a) of Figure 15.1.

### (a) Limits to growth stock and flow model

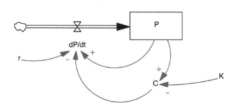

### (b) Limits to growth differential equation model

$$\frac{dP}{dt} = r\,P\,(1 - C) \quad (1) \qquad\qquad C = P/K \quad (2) \qquad\qquad r = 0.15 \quad (3)$$

$$K = 100{,}000 \quad (4) \qquad\qquad P_{INIT} = 100 \quad (5)$$

**FIGURE 15.1** Stock and flow model for limits to growth

There are a number of points to be made in relation to this model.

- There is one stock ($P$) which models the population (for example, the population of a growing city). This stock has one inflow (the net flow), as shown by Equation (1).

- The net flow (1) has two interesting elements. The first is the product $rP$ which represents the number added to the population in the absence of any limiting factor. The second term $(1 - C)$ can be viewed as a moderator on the net flow. For example, when $C = 1$, the net flow is *switched off* and no further growth can happen in the system. Equation (2) in Figure 15.1

determines the value of $C$, and this is the current population divided by the population limit $K$ (4), where $K$ is often referred to as the carrying capacity.

- There are a number of constants in the model, and these are arbitrary choices. They include the growth rate $r$ (3), the carrying capacity $K$ (4), and the initial population value $P_{INIT}$ (5).

- Overall, there are two feedback loops in the model. The first loop can be traced from $P$ to $dP/dt$ and back to $P$, and this is a *reinforcing loop* that drives growth in the system. The second loop also starts at $P$, then to $C$, $dP/dt$ before connecting back to $P$. This loop, because of the single negative polarity sign, is a *balancing loop*, and therefore counteracts the system growth. This makes sense, as the carrying capacity $K$ ultimately impacts the growth in $P$.

We will now take this stock and flow model and implement it using deSolve.

---

## 15.3   deSolve

R's deSolve package solves initial value problems written as ordinary differential equations (ODEs), differential algebraic equations (DAEs), and partial differential equations (PDEs) (Soetaert et al., 2010). We will make use of the function ode(), which solves a system of ordinary differential equations (i.e., a system of stock and flow equations). It takes the following arguments.

```
ode(y, times, func, parms,
method = c("lsoda", "lsode", "lsodes", "lsodar", "vode", "daspk",
           "euler", "rk4", "ode23", "ode45", "radau",
           "bdf", "bdf_d", "adams", "impAdams",
           "impAdams_d", "iteration"),...)
```

These arguments are summarized, and this information can be retrieved in full using the command ?deSolve::ode.

1. **y**: The initial (state) values for the ODE system, a vector.

2. **times**: The time sequence for which output is wanted; the first value of times must be the initial time.

3. **func**: The user-defined R function that computes the values of the derivatives in the ODE system (the model definition) at time t. Defined as: func <- function(t, y, parms,...) where t is the current time point of the integration, and y is the current estimate of the ODE system variables. The derivatives must be specified in the same order as the state variables y. The return value of func should be a

list whose first element is a vector containing the derivatives of y with respect to time, and whose next elements are (optional) global values to be recorded.

4. **params**: Parameters passed to the function `func`.

5. **method**: A string to indicate the numerical integration method, for example, "euler", "rk4", "ode23", "ode45".

6. **returns**: A matrix of S3 class `deSolve` with up to as many rows as elements in times and as many columns as elements in y plus the number of "global" values returned. This can be easily converted to a data frame object using the function `data.frame()`.

We will now implement the limits to growth model using `deSolve`. First, we include the necessary packages, and `dplyr`, `tidyr`, and `ggplot2` to ensure that we can store the results in a `tibble` and then display the graphs over time with `ggplot2`. The library `purrr` is used to run sensitivity analysis, and `GGally` allows for multiple plots to be combined into a single plot.

```
library(deSolve)
library(dplyr)
library(ggplot2)
library(tidyr)
library(purrr)
library(ggpubr)
```

Before exploring the function code, we describe an additional R function that is used for convenience as part of the function referenced by the argument `func`. The R function `with()` evaluates a expression in an environment constructed from data. It allows us to store data in a list, and then operate on that data using only the list element name. We employ this to make the model equations more readable, as we do not have to use the conventional subsetting notation for vectors that was covered in Chapter 3.

To show how `with()` works, we first consider the following code, where we have two vectors v1 and v2 and we want to add the elements together by directly accessing the named elements. This is one solution.

```
v1 <- c(a=10)
v2 <- c(b=20,c=30)
ans1 <- v1["a"] + v2["b"] + v2["c"]
ans1
#>  a
#> 60
```

A second solution is to combine the two vectors into one, convert the results to a list, and then use the list operator to add the elements.

```
l1 <- as.list(c(v1,v2))
ans2 <- l1$a + l1$b + l1$c
ans2
#> [1] 60
```

A third approach is to pass the list into the with() function, and then use the facility within this function to access the elements directly by their names. The return statement is used to pass back the result of this calculation, which is stored in the variable ans3.

```
ans3 <- with(as.list(c(v1,v2)),{
  return (a+b+c)
})
ans3
#> [1] 60
```

This use of with() can now be seen in our implemention for the limits to growth model. The function, named ltg, has the list of arguments that ode will expect (time, stocks, and auxs in this case, where we have renamed the third parameter to a term used more widely in system dynamics). The named arguments from stocks and auxs are combined into a list, and passed to the with() function, and the model equations for C and dP_dt are specified. The derivative is returned as the *first list element*, and other variables which we want to record in the final output are added as additional list elements.

```
ltg <- function(time, stocks, auxs){
  with(as.list(c(stocks, auxs)),{
    C <- P/K                # Eq (2)
    dP_dt <- r*P*(1-C)      # Eq (1)
    return (list(c(dP_dt),
                 r=r,
                 K=K,
                 C=C,
                 Flow=dP_dt))
  })
}
```

Our next step is to configure the remaining vectors for the simulation. We define the simtime vector to indicate the simulation's start and finish time, and the intervening steps we want displayed (in this case, we want results for each step of 0.25). The vector stocks contains the list of stocks and their initial values. Note that the stock name is P, which must be the same as the variable we use within the function ltg. We specify the vector auxs, and this contains the auxiliaries (or parameters) of the model. Again, the name choice is important, as the variables r and K are referenced within the function ltg.

```
simtime <- seq(0,100,by=0.25)
stocks  <- c(P=100)                    # Eq (5)
auxs    <- c(r=0.15,K=100000)   # Eq (3) and Eq (4)
```

We are now ready to run the simulation, by calling the function ode(). The function takes five arguments:

- stocks, the stocks in the model, along with their initial values.
- simtime, the time sequence for the output.
- ltg, the function that contains the model to be simulated, and it calculates the derivatives.
- auxs, the model auxiliaries.
- "euler", which selects the numerical integration method for the problem.

```
res <- ode(y=stocks,
           times=simtime,
           func = ltg,
           parms=auxs,
           method="euler") %>%
       data.frame() %>%
       dplyr::as_tibble()
res
#> # A tibble: 401 x 6
#>      time     P     r       K         C Flow
#>     <dbl> <dbl> <dbl>   <dbl>     <dbl> <dbl>
#>  1  0       100  0.15 100000 0.001      15.0
#>  2  0.25   104.  0.15 100000 0.00104    15.5
#>  3  0.5    108.  0.15 100000 0.00108    16.1
#>  4  0.75   112.  0.15 100000 0.00112    16.7
#>  5  1      116.  0.15 100000 0.00116    17.4
#>  6  1.25   120.  0.15 100000 0.00120    18.0
#>  7  1.5    125.  0.15 100000 0.00125    18.7
#>  8  1.75   129.  0.15 100000 0.00129    19.4
#>  9  2      134.  0.15 100000 0.00134    20.1
#> 10  2.25   139.  0.15 100000 0.00139    20.9
#> # ... with 391 more rows
```

As ode() returns a matrix, we transform this to a data frame, and then onto a tibble, and the first ten results are displayed. Note that the variable P is increasing, as expected, and the crowding variable C, while initially small, is also starting to increase. This will eventually reach 1, and at that point the derivative equation will be set to zero, so all growth will cease.

Both the stock and the flow are shown below, and they display the impact of the crowding variable $C$ on the flow, and hence the stock. These are visualized in Figure 15.2.

```
res_long <- res %>%
            dplyr::select(time,C,P,Flow) %>%
            tidyr::pivot_longer(names_to = "Variable",
                                values_to = "Value",
                                -time)
```

```
ggplot(res_long,aes(x=time,y=Value,color=Variable)) +
  geom_line() + facet_wrap(~Variable,scales = "free")+
  theme(legend.position = "top")
```

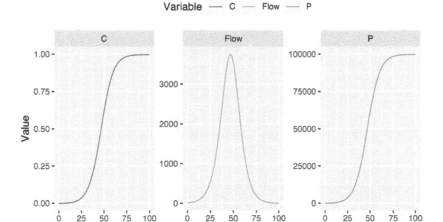

**FIGURE 15.2** Model outputs following the simulation

Following this first deSolve model, we will now explore two models of infectious disease transmission, the SIR and the SIRH models.

## 15.4   Susceptible-infected-recovered model

The stock and flow structure for the susceptible-infected-recovered (SIR), and model equations, are shown in Figure 15.3. This is a classic model for simulating disease dynamics in a population, where it is assumed that everyone can contact everyone else. The model assumption is that the total population remains constant, as there are no births, deaths, or migrations. This is a reasonable assumption if we consider a relatively short duration for the outbreak. Note that some SIR models use the term *removed* instead of *recovered*.

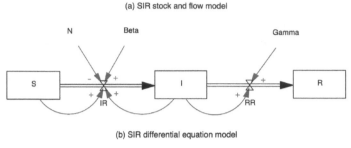

(a) SIR stock and flow model

(b) SIR differential equation model

$$\frac{dS}{dt} = -IR \quad (6) \qquad \frac{dI}{dt} = IR - RR \quad (7) \qquad \frac{dR}{dt} = RR \quad (8)$$

$$IR = \beta I \frac{S}{N} \quad (9) \qquad RR = I\gamma \quad (10) \qquad N = 10{,}000 \quad (11)$$

$$\gamma = 0.25 \quad (12) \qquad \beta = 1.0 \quad (13) \qquad S_{INIT} = 9999 \quad (14)$$

$$I_{INIT} = 1 \quad (15) \qquad R_{INIT} = 0 \quad (16)$$

**FIGURE 15.3** Susceptible-infected-recovered stock and flow model

The model divides the population into three stocks (often called compartments in disease modelling), where each stock captures a different disease state. These stocks are:

- Suceptible (6), where people residing in this stock have no prior immunity to a pathogen and so may become infected. This stock has one outflow (9).
- Infected (7), where people have been infected and can also go on to infect others. People do not reside in the infected stock indefinitely, and after a duration, will exit and move to the recovered stock. This stock has one inflow (9) and one outflow (10).
- Recovered (8), where people cannot infect others, nor can then be infected themselves. This stock has one inflow (10).

The infection rate $IR$ (9) is a key driver of model behavior, and it has an intuitive structure, and comprises a number of terms.

- First, the effective contact rate $\beta$ (13), which represents a contact that is sufficient to lead to transmission if it occurs between an infectious and susceptible person (Vynnycky and White, 2010). For example, if an infectious person met 10 susceptible people in one day, and passed on a virus with a 10% chance, then the $\beta$ value would be 1.0.

- Second, the number of effective contacts that infected people generate, which is $\beta I$, for example, if there were 10 infected people, then there would be 10 effective contacts when $\beta = 1.0$.

- Third, we calculate the chance that these contacts will be made with a susceptible person, which is $S/N$.

The recovery rate $RR$ (10) ensures that the infected stock depletes, as in reality, people who become infected are typically only infectious for a certain duration. A common way to model a delay duration in a stock and flow model is to invert the delay to form a fraction, and decrease the stock by this fraction for each time step of the simulation. Therefore, $\gamma = 0.25$ (12) models an average infectious delay of four days.

We now implement the SIR model equations in R, using deSolve. The function containing the model equations is shown below. The aim of the function is to calculate the flow variables (derivatives) for the three stocks.

```
sir <- function(time, stocks, auxs){
  with(as.list(c(stocks, auxs)),{
    N      <- 10000        # Eq (11)
    IR     <- beta*I*S/N   # Eq (9)
    RR     <- gamma*I      # Eq (10)
    dS_dt <- -IR           # Eq (6)
    dI_dt <- IR - RR       # Eq (7)
    dR_dt <- RR            # Eq (8)
    return (list(c(dS_dt,dI_dt,dR_dt),
                 Beta=beta,
                 Gamma=gamma,
                 Infections=IR,
                 Recovering=RR))
  })
}
```

Before running the model, the time, stocks, and auxiliaries vectors are created.

```
simtime <- seq(0,50,by=0.25)
stocks  <- c(S=9999,I=1,R=0)     # Eq (14), Eq (15), and Eq (16)
auxs    <- c(gamma=0.25,beta=1) # Eq (12) and Eq (13)
```

With these three vectors in place, the simulation can be run by calling the deSolve function ode. This simulation runs based on the specified auxiliaries and initial stock values, and varying any of these would yield a different result. The call to ode() returns a matrix, which is then converted to a tibble. The results are shown, where each row is the simulation output for a time step, and each column is a model variable.

```
res <- ode(y=stocks,
           times=simtime,
           func = sir,
           parms=auxs,
           method="euler") %>%
```

```
        data.frame() %>%
        dplyr::as_tibble()
res
#> # A tibble: 201 x 8
#>      time     S     I      R Beta Gamma Infections Recovering
#>     <dbl> <dbl> <dbl>  <dbl> <dbl> <dbl>      <dbl>      <dbl>
#>  1  0      9999  1     0        1  0.25       1.00       0.25
#>  2  0.25  9999.  1.19 0.0625    1  0.25       1.19       0.297
#>  3  0.5   9998.  1.41 0.137     1  0.25       1.41       0.353
#>  4  0.75  9998.  1.67 0.225     1  0.25       1.67       0.419
#>  5  1     9998.  1.99 0.329     1  0.25       1.99       0.497
#>  6  1.25  9997.  2.36 0.454     1  0.25       2.36       0.590
#>  7  1.5   9997.  2.80 0.601     1  0.25       2.80       0.701
#>  8  1.75  9996.  3.33 0.777     1  0.25       3.33       0.832
#>  9  2     9995.  3.95 0.985     1  0.25       3.95       0.988
#> 10  2.25  9994.  4.69 1.23      1  0.25       4.69       1.17
#> # ... with 191 more rows
```

In order to visualize the results, using `pivot_longer()` from `tidyr` we create
two new tibbles `flows_piv` and `stocks_piv`, which will allow us to plot the
flows on one graph, and the stocks on another.

```
flows_piv <- res %>%
             dplyr::select(time,Infections:Recovering) %>%
             tidyr::pivot_longer(names_to="Flow",
                                 values_to = "Value",
                                 -time) %>%
             dplyr::mutate(Flow=factor(Flow,
                                  levels = c("Infections",
                                             "Recovering")))


stocks_piv <- res %>%
              dplyr::select(time,S:R) %>%
              tidyr::pivot_longer(names_to="Stock",
                                  values_to = "Value",
                                  -time) %>%
              dplyr::mutate(Stock=factor(Stock,
                                   levels = c("S","I","R")))
```

Two separate plots are created, `p1` and `p2`, and are display in Figure 15.4.
These are then combined using the package `ggpubr` which contains the function
`ggarrange()` that allows a number of plots to be combined.

```
p1 <- ggplot(flows_piv,aes(x=time,y=Value,color=Flow))+geom_line()+
      theme(legend.position = "top")+
      labs(x="Day",y="Flows")
```

```
p2 <- ggplot(stocks_piv,aes(x=time,y=Value,color=Stock))+geom_line()+
    theme(legend.position = "top")+
    labs(x="Day",y="Stocks")

g1 <- ggarrange(p1,p2,nrow = 2)
g1
```

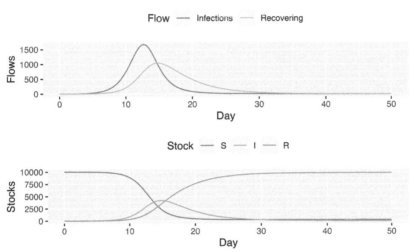

**FIGURE 15.4** Stock and flow outputs from the model

The first plot shows the model flows, and for example, when the inflow exceeds the outflow, we can see that the infected stock (second plot) rises. At all times the three stocks will sum to 10,000 as all flows are kept within the three stocks.

## 15.5 Susceptible-infected-recovered-hospital model

A benefit of the system dynamics approach is that it facilitates extending the model boundary to explore additional impacts of the system under study. In the case of the SIR model, an important question to consider is: are there possible downstream effects resulting from the spread of a novel pathogen?

In this case, we propose that there are impacts. People who are infected by the pathogen may experience a medical event that requires that they must be admitted to the hospital for a certain duration before they fully recover. We can model this in a simple way and also model possible countermeasures to reduce the impact of the pathogen spread. The new structure and equations are shown in Figure 15.5. Note the the new structure is a simplified model;

its purpose here is to explore the effect of different countermeasures on the hospital occupancy numbers.

(a) SIRH stock and flow model

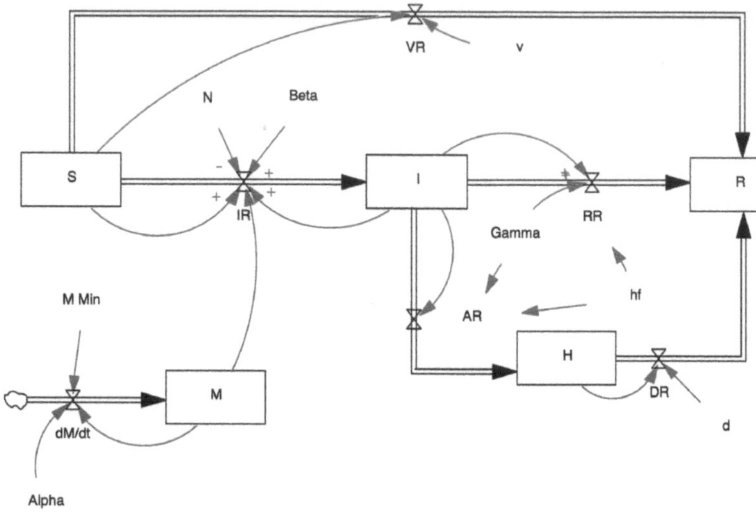

(b) SIRH differential equation model

$$\frac{dS}{dt} = -IR - VR \quad (17) \qquad \frac{dI}{dt} = IR - RR - AR \quad (18) \qquad \frac{dH}{dt} = AR - DR \quad (19)$$

$$\frac{dR}{dt} = RR + DR + VR \quad (20) \qquad \frac{dM}{dt} = \alpha \, (M_{min} - M) \quad (21) \qquad IR = \beta \, M \, I \, \frac{S}{N} \quad (22)$$

$$VR = v \, S \quad (23) \qquad AR = h_f I \, \gamma \quad (24) \qquad RR = (1 - h_f) I \, \gamma \quad (25)$$

$$DR = H \, d \quad (26) \qquad \beta = 1.0 \quad (27) \qquad \gamma = 0.25 \quad (28)$$

$$v = 0.1 \quad (29) \qquad h_f = 0.1 \quad (30) \qquad N = 10{,}000 \quad (31)$$

$$d = 0.10 \quad (32) \qquad \alpha = 0.5 \quad (33) \qquad S_{INIT} = 9999 \quad (34)$$

$$I_{INIT} = 1 \quad (35) \qquad H_{INIT} = 0 \quad (36) \qquad R_{INIT} = 0 \quad (37)$$

$$M_{INIT} = 1 \quad (38) \qquad M_{MIN} = 0.3 \quad (39)$$

**FIGURE 15.5** Susceptible-infected-recovered-hospital stock and flow model

The SIRH structure has a number of features:

- A new stock $H$ (19) is introduced, which models the number of people in the hospital. The assumption here is that when people are no longer infectious, the proportion $h_f$ (30) will be diverted to the hospital via the flow $AR$ (24), which models the hospital admission rate. A simplification here is that the flow $AR$ is not constrained by a lack of hospital resources, which is unrealistic for an operational model, but appropriate here to explore some interesting dynamics and investigate a number of scenarios. People remain in the hospital based on the delay fraction $d$ (32), after which they flow, via $DR$ (26), to the recovered stock (20).

- A medical countermeasure is introduced to the model, and this is vaccination. This is achieved by connecting the flow $VR$ from $S$ to $R$, via a fractional multiplier $v$ (29). So what are the assumptions underlying this flow? It models the rate at which people are vaccinated, and it also assumes that this process is fully effective. While this assumption is a simplification, it is reasonable for an exploratory model.

- A non-pharmaceutical intervention is added to the model in the form of a mobility multiplier $M$ (21). The behavior of this mobility stock is governed by three parameters: $M_{INIT}$, $M_{MIN}$, and $\alpha$. $M_{INIT}$ specifies the initial value of the mobility multiplier, and is set to 1, which means that initially, it does not impact the flow $IR$. $M_{MIN}$ specifies the absolute minimum for $M$, and $\alpha$ determines how quickly $M$ reaches this value. For example, if $\alpha = 0$, then $M$ will never change, and so this value of $\alpha$ models the scenario where no mobility restrictions are introduced. Overall, the use of a stock to model the multiplier $M$ is effective, as it allows us to specify time-varying behavior for the variable in a simple and robust way.

Note that the equations for $S$ (17), $I$ (18), and $R$ (20) have changed from the original SIR model, and this is due to the addition of flows to model hospitalization ($AR$,$DR$) and medical countermeasures ($VR$). The flow $IR$ (22) has also been modified to take into account mobility reductions, via the multiplier $M$ (21), which is implemented as a stock structure.

The function sirh contains the model equations, and returns the five key derivatives for the ode() function. It also returns global variables that are added to the result, including the variable CheckSum which is a conservation test to ensure that the sum of all stocks at any point in time should equal the total population $N$.

```
sirh <- function(time, stocks, auxs){
  with(as.list(c(stocks, auxs)),{
    N <- 10000              # Eq (31)
    IR <- beta*I*S/N*M      # Eq (22)
    VR <- v*S               # Eq (23)
    AR <- hf*gamma*I        # Eq (24)
    RR <- (1-hf)*gamma*I    # Eq (25)
    DR <- d*H               # Eq (26)


    dS_dt  <- -IR -VR        # Eq (17)
    dI_dt  <-  IR - AR - RR  # Eq (18)
    dH_dt  <-  AR - DR       # Eq (19)
```

```
    dR_dt  <-  RR + DR + VR  # Eq (20)
    dM_dt  <-  (M_min - M) *
               alpha         # Eq (21)
    return (list(c(dS_dt,dI_dt,dH_dt,dR_dt,dM_dt),
                 Beta=beta,
                 Gamma=gamma,
                 HF=hf,
                 V=v,
                 Alpha=alpha,
                 M_Min=M_min,
                 Infections=IR,
                 Recovering=RR,
                 Vaccinated=VR,
                 Hospitalized=AR,
                 Discharged=DR,
                 CheckSum=S + I + R + H))
  })
}
```

For this model, an additional goal is to run many scenarios and explore the results. This will facilitate policy exploration, and allow us to view the combined impact of medical and non-pharmaceutical countermeasures.

## 15.6  Policy exploration of the SIRH model using sensitivity analysis

With the model specified in the function sirh(), we can now add another function to support sensitivity analysis, where we can run the model for a set of parameters and then explore the results. Sensitivity analysis is a widely used technique in system dynamics, as it generates many different simulations, the results of which can be analyzed. The new function is called run_scenario(), which:

- Takes in a set of model arguments (all of which have default values).
- Populates the vector auxs with the auxiliary values.
- Calls the function ode() to run the model, converts the results to a tibble, and adds a column (RunID) to record the run identifier.

```
run_scenario <- function(run_id=1,
                         stocks=c(S=9999, # Eq (34)
                                  I=1,    # Eq (35)
                                  H=0,    # Eq (36)
                                  R=0,    # Eq (37)
```

```
                            M=1          # Eq (38)
                          ),
                    simtime=seq(0,50,by=0.25),
                    beta=1,          # Eq (27)
                    gamma=0.25,      # Eq (28)
                    alpha=0.5,       # Eq (23)
                    M_min=0.3,       # Eq (39)
                    d=0.1,           # Eq (32)
                    hf=0.1,          # Eq (30)
                    v=0.1            # Eq (29)
                  ){
  auxs <- c(beta=beta,
            gamma=gamma,
            alpha=alpha,
            M_min=M_min,
            d=d,
            hf=hf,
            v=v)

  res <- ode(y=stocks,
             times=simtime,
             func = sirh,
             parms=auxs,
             method="euler") %>%
        data.frame() %>%
        dplyr::as_tibble() %>%
        dplyr::mutate(RunID=as.integer(run_id)) %>%
        dplyr::select(RunID, everything())
}
```

The advantage of creating the function run_scenario() is that we can now call this for a range of parameter values. Here, we are going to sample two policy variables:

- $\alpha$, which models the speed of mobility restriction implementations. For example, a higher value of $\alpha$ would mean that the population responds quickly to the request for social mobility reductions. In our simulations $0 \leq \alpha \leq 0.20$.

- $v$, which models the speed of vaccination. A higher value of $v$ means that people transfer more quickly from $S$ to $R$, and therefore the burden on the hospital sector should be reduced. In our simulations $0 \leq v \leq 0.05$, which indicates that the minimum vaccination duration is 20 days (1/0.05).

When we reflect on the role of these two parameters, we would expect that a combination of the highest values for both $\alpha$ and $v$ should result in the best outcome. Alternatively, the lowest values for $\alpha$ and $v$ should generate the worst outcome, as there will be no vaccination ($v = 0$), and no mobility restrictions ($\alpha = 0$).

We will sample the parameter space evenly with 50 equally spaced samples for each parameter, using the `length.out` argument with the `seq()` function. The function `expand.grid()` generates the range of combinations of these parameter values. This will result in 2,500 simulations.

```
alpha_vals <- seq(0,.20,length.out=50)
vacc_vals  <- seq(0,0.05,length.out=50)
sim_inputs <- expand.grid(alpha_vals,vacc_vals)
summary(sim_inputs)
#>      Var1              Var2
#> Min.    :0.000   Min.    :0.0000
#> 1st Qu.:0.049   1st Qu.:0.0122
#> Median :0.100   Median :0.0250
#> Mean    :0.100   Mean    :0.0250
#> 3rd Qu.:0.151   3rd Qu.:0.0378
#> Max.    :0.200   Max.    :0.0500
```

The sensitivity analysis loop can then be formulated using the `map2()` function from `purrr`, where the two inputs are the values for $\alpha$ and $v$. The `map2()` function returns a list of simulation result, where each list element contains the full simulation results for an individual run. The function `bind_rows()` then combines all of the list elements into one large data frame, and the results are stored in the tibble `sim_res`. Note that in contrast to the agent-based simulations in Chapter 14, we chose not to use the parallel version of `map2()`, because the overall runtime associated with the SIRH model is not as high as the agent-based model.

```
run_id <- 1

sim_res <- map2(sim_inputs[,1],
                sim_inputs[,2],~{
  res <- run_scenario(run_id   = run_id,
                      alpha    = .x,
                      v        = .y,
                      simtime  = seq(0,75,by=0.25),
                      hf       = 0.10)
  run_id <<- run_id + 1
  res
}) %>% dplyr::bind_rows()
sim_res
#> # A tibble: 752,500 x 19
#>    RunID  time      S      I      H      R      M   Beta  Gamma     HF
#>    <int> <dbl>  <dbl>  <dbl>  <dbl>  <dbl>  <dbl>  <dbl>  <dbl>  <dbl>
```

```
#>  1      1  0     9999   1     0        0          1      1  0.25  0.1
#>  2      1  0.25  9999.  1.19  0.00625  0.0562     1      1  0.25  0.1
#>  3      1  0.5   9998.  1.41  0.0135   0.123      1      1  0.25  0.1
#>  4      1  0.75  9998.  1.67  0.0220   0.203      1      1  0.25  0.1
#>  5      1  1     9998.  1.99  0.0319   0.298      1      1  0.25  0.1
#>  6      1  1.25  9997.  2.30  0.0435   0.410      1      1  0.25  0.1
#>  7      1  1.5   9997.  2.80  0.0572   0.544      1      1  0.25  0.1
#>  8      1  1.75  9996.  3.33  0.0733   0.703      1      1  0.25  0.1
#>  9      1  2     9995.  3.95  0.0923   0.892      1      1  0.25  0.1
#> 10      1  2.25  9994.  4.69  0.115    1.12       1      1  0.25  0.1
#> # ... with 752,490 more rows, and 9 more variables: V <dbl>,
#> #   Alpha <dbl>, M_Min <dbl>, Infections <dbl>,
#> #   Recovering <dbl>, Vaccinated <dbl>, Hospitalized <dbl>,
#> #   Discharged <dbl>, CheckSum <dbl>
```

In addition to the full set of results, information on the 95th and 5th quantiles can be calculated, when the results are grouped by the variable `time`.

```
time_h <- sim_res %>%
  dplyr::group_by(time) %>%
  dplyr::summarize(MeanH=mean(H),
                   Q95=quantile(H,0.95),
                   Q05=quantile(H,0.05))

time_h
#> # A tibble: 301 x 4
#>     time   MeanH      Q95      Q05
#>    <dbl>   <dbl>    <dbl>    <dbl>
#>  1  0     0        0        0
#>  2  0.25  0.00625  0.00625  0.00625
#>  3  0.5   0.0135   0.0135   0.0135
#>  4  0.75  0.0219   0.0220   0.0219
#>  5  1     0.0317   0.0319   0.0316
#>  6  1.25  0.0430   0.0434   0.0426
#>  7  1.5   0.0560   0.0569   0.0551
#>  8  1.75  0.0709   0.0726   0.0692
#>  9  2     0.0880   0.0911   0.0851
#> 10  2.25  0.107    0.113    0.103
#> # ... with 291 more rows
```

Three plots, shown in Figure 15.6, are generated and displayed together using the function `ggarrange()`. The plots include p3, which shows the infected stock for all runs, p4 that shows the hospital stock for all runs, and p5 which displays the mean and quantiles for the hospital stock. Interestingly, from the plot, the lag between infection and hospital peaks can be observed.

```
p3 <- ggplot(sim_res,aes(x=time,y=Infections,color=RunID,group=RunID))+
   geom_line()+
   scale_color_gradientn(colors=rainbow(14))+
   theme(legend.position = "none")+
   labs(title="Infections (flow)")+
   theme(title = element_text(size=9))

p4 <- ggplot(sim_res,aes(x=time,y=H,color=RunID,group=RunID))+
   geom_line()+
   scale_color_gradientn(colors=rainbow(14))+
   theme(legend.position = "none")+
   labs(title="People in hospital (stock)")+
   theme(title = element_text(size=9))

p5 <- ggplot(time_h,aes(x=time,y=MeanH))+geom_line()+
       geom_ribbon(aes(x=time,ymin=Q05,ymax=Q95),
                    alpha=0.4,fill="steelblue2")+
       labs(title="90% quantiles for people in hospital")+
       theme(title = element_text(size=9))

g2 <- ggarrange(p3,p4,p5,nrow = 3)
g2
```

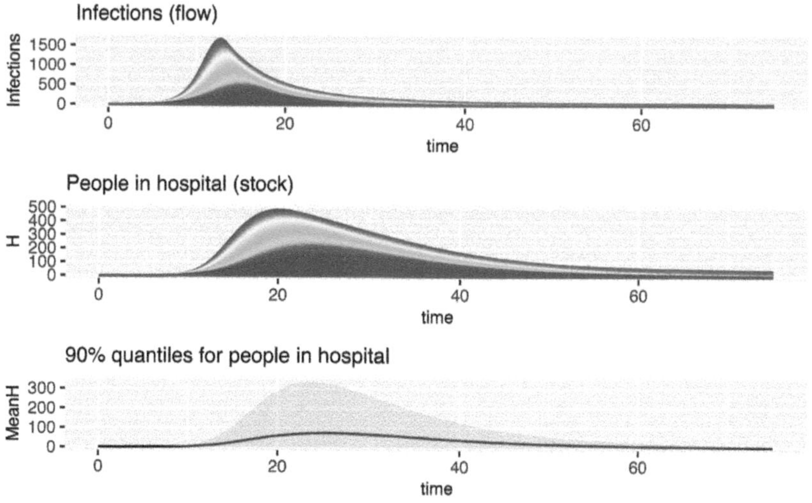

**FIGURE 15.6** Exploring the model results

Another type of plot can be generated, which focuses on a scatterplot for the parameters $\alpha$ and $v$ and uses information relating to the maximum number of people in the hospital for each simulation run. The `dplyr` code to generate this data is shown below, where we can extract the two parameter values, alongside the maximum size of the hospital stock. This hospital stock value can be used as an indicator for the overall pressure on the health system.

```
max_h <- sim_res %>%
            dplyr::group_by(RunID) %>%
            dplyr::summarize(MH=max(H),
                             V=first(V),
                             Alpha=first(Alpha))
```

If we consider the range of simulations, it is valuable to think about what we might expect to see from the results, and to explore this, we first present the results in descending order of MH.

As you might expect, the highest numbers in the hospital system are recorded when the two parameter values are lowest. Specifically, when mobility is not reduced ($\alpha = 0$) and there is no vaccination protection ($v = 0$).

```
max_h %>% dplyr::arrange(desc(MH)) %>% dplyr::slice(1:3)
#> # A tibble: 3 x 4
#>    RunID    MH        V   Alpha
#>    <int> <dbl>    <dbl>   <dbl>
#> 1      1  487. 0            0
#> 2      2  480. 0            0.00408
#> 3     51  479. 0.00102 0
```

On the other hand, when we explore the values by ascending order of MH, we can see that the maximum number in the hospital system is lowest when the parameter values are at their maximum.

```
max_h %>% dplyr::arrange(MH) %>% dplyr::slice(1:3)
#> # A tibble: 3 x 4
#>    RunID    MH        V Alpha
#>    <int> <dbl>    <dbl> <dbl>
#> 1   2500  1.68 0.05     0.2
#> 2   2450  1.72 0.0490   0.2
#> 3   2499  1.73 0.05     0.196
```

While it is informative to view the data, visualization is valuable when exploring the overall policy space, and patterns are evident when the data is plotted. We have two plots to explore the policy space and interaction between the parameters, which are combined into the variable g3:

- First, using a conventional plot (p6) we color and size the points based on the variable MH, and we show these points on a color gradient where the lowest values are colored blue, and the highest values are displayed in red.

- Second, using the function geom_contour_filled() we generate an informative contour plot (right hand side of Figure 15.7) that shows the areas of similar height, with the highest range, from 450 to 500, shown in yellow. This diagram is informative as it indicates the common range of outcomes for different parameter combinations. For example, the darkest color indicates the lowest range between 0 and 50 which shows the combinations of parameter values that yield the best results.

```
p6 <- ggplot(max_h,aes(x=Alpha,y=V,color=MH,size=MH))+geom_point()+
  scale_color_gradient(low="blue", high="red")+
  theme(legend.position = "none")+
  labs(title=paste0("Parameter Analysis"))+
  labs(subtitle=paste0("Max peak = ",
                       round(max(max_h$MH),0),
                       " at point (0,0)"))+
  theme(title = element_text(size=9))



p7 <- ggplot(max_h,aes(x=Alpha,y=V,z=MH))+geom_contour_filled()+
  theme(legend.position = "none")+
  labs(title=paste0("Contour plot"))+
  labs(subtitle=paste0("Yellow band range (450,500]"))+
  theme(title = element_text(size=9))

g3 <- ggarrange(p6,p7,nrow = 1)
g3
```

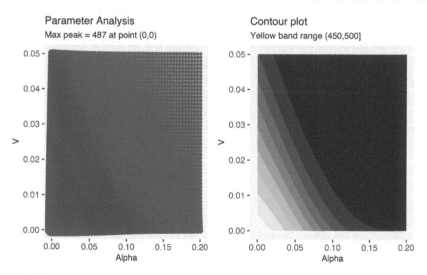

**FIGURE 15.7** Policy exploration of the SIRH model using sensitivity analysis

That concludes the overall policy exploration example. While the model is simple in terms of a small number of stocks, and two policies, it does provide an insight into the type of analysis that can be performed using system dynamics models and the deSolve package, in tandem with the tools of the tidyverse, which include purrr, ggplot2, dplyr, and tidyr.

## 15.7   Summary of R functions from Chapter 15

A list of key functions introduced in this chapter is now shown.

| Function | Description |
| --- | --- |
| ode() | Solves a system of ODEs (deSolve). |
| with() | Evaluates an expression in an environment constructed from data. |

# Bibliography

(2008). *Exploratory Data Analysis*. The Concise Encyclopedia of Statistics. Springer, New York.

Bache, S. M. and Wickham, H. (2014). *magrittr: A forward-pipe operator for R*.

Barabási, A. (2016). Network science.

Barabási, A.-L. and Albert, R. (1999). Emergence of scaling in random networks. *Science*, 286(5439):509–512.

Bass, F. M. (2004). Comments on "a new product growth for model consumer durables the bass model". *Management Science*, 50(12_supplement):1833–1840.

Baumer, B. S., Kaplan, D. T., and Horton, N. J. (2021). *Modern Data Science with R*. CRC Press.

Chambers, J. M. (2008). *Software for Data Analysis: Programming with R*, volume 2. Springer.

Chambers, J. M. (2017). *Extending R*. Chapman & Hall/CRC.

Crawley, M. J. (2015). *Statistics: An Introduction Using R (2nd Edition)*. John Wiley & Sons, Chichester.

Epstein, J. M. (2012). Generative social science. In *Generative Social Science*. Princeton University Press.

Field, A., Miles, J., and Field, Z. (2012). *Discovering Statistics using R*. Sage Publications Ltd, London, UK.

Ford, D. N. (2019). A system dynamics glossary. *System Dynamics Review*, 35(4):369–379.

Forrester, J. W. (1961). *Industrial Dynamics*. Productivity Press, Portland, Oregon.

Harrison Jr, D. and Rubinfeld, D. L. (1978). Hedonic housing prices and the demand for clean air. *Journal of Environmental Economics and Management*, 5(1):81–102.

Hillier, F. S. and Lieberman, G. J. (2019). *Introduction to Operations Research.* McGraw Hill, eleventh edition.

Hoel, P. G. and Jessen, R. J. (1971). *Basic Statistics for Business and Economics.* John Wiley & Sons, New York.

Hyndman, R. J. and Athanasopoulos, G. (2021). *Forecasting: Principles and Practice.* OTexts.

James, G., Witten, D., Hastie, T., Tibshirani, R., et al. (2013). *An Introduction to Statistical Learning*, volume 112. Springer.

Knaflic, C. N. (2015). *Storytelling with Data: A Data Visualization Guide for Business Professionals.* John Wiley & Sons.

Kuhn, M. and Silge, J. (2022). *Tidy Modeling with R.* O'Reilly Media, Inc.

Lander, J. P. (2017). *R for everyone: Advanced Analytics and Graphics.* Pearson Education, second edition.

Lantz, B. (2019). *Machine Learning with R: Expert Techniques for Predictive Modeling.* Packt Publishing Ltd.

Matloff, N. (2011). *The Art of R Programming: A Tour of Statistical Software Design.* No Starch Press.

Montgomery, D. C. (2007). *Introduction to Statistical Quality Control.* John Wiley & Sons.

Morecroft, J. D. (2015). *Strategic Modelling and Business Dynamics: A Feedback Systems Approach.* John Wiley & Sons.

Newman, M. (2018). *Networks.* Oxford University Press.

Richardson, G. P. (1991). *Feedback Thought in Social Science and System Theory.* Pegasus Communications.

Soetaert, K., Petzoldt, T., and Setzer, R. W. (2010). Solving Differential Equations in R: Package desolve. *Journal of Statistical Software*, 33(9):1–25.

Sterman, J. (2000). *Business Dynamics: Systems Thinking and Modeling for a Complex World.* McGraw-Hill.

Taha, H. A. (1992). *Operations Research: An Introduction.* Macmillan Publishing Company, fifth edition.

Vynnycky, E. and White, R. (2010). *An Introduction to Infectious Disease Modelling.* OUP Oxford.

Warren, K. (2008). *Strategic Management Dynamics.* John Wiley & Sons.

Watts, D. J. and Strogatz, S. H. (1998). Collective dynamics of 'small-world' networks. *Nature*, 393(6684):440–442.

Wickham, H. (2014). Tidy data. *Journal of Statistical Software*, 59(10):1–23.

Wickham, H. (2016). *ggplot2: Elegant Graphics for Data Analysis*. Springer New York, NY.

Wickham, H. (2019). *Advanced R*. CRC Press.

Wickham, H. (2021). *Mastering Shiny*. O'Reilly Media, Inc.

Wickham, H., Çetinkaya-Rundel, M., and Grolemund, G. (2023). *R for Data Science*. O'Reilly Media, Inc.

Wickham, H. and Grolemund, G. (2016). *R for Data Science: Import, Tidy, Transform, Visualize, and Model Data*. O'Reilly Media, Inc.

Wilensky, U. and Rand, W. (2015). *An Introduction to Agent-Based Modeling: Modeling Natural, Social, and Engineered Complex Systems with NetLogo*. Mit Press.

Xie, Y., Allaire, J. J., and Grolemund, G. (2018). *R Markdown: The Definitive Guide*. CRC Press.

# Index

Milton Keynes UK
Ingram Content Group UK Ltd.
UKHW022036141024
449569UK00014B/629